Tissue Engineering

W. W. Minuth, R. Strehl, K. Schumacher

Further Titles of Interest

Novartis Foundation Symposium

**Tissue Engineering
of Cartilage and Bone –
No. 249**

2003
ISBN 0-470-84481-7

Alan Doyle, J. Bryan Griffiths (Eds.)

**Cell and Tissue Culture
for Medical Research**

2000
ISBN 0-471-85213-9

R. Ian Freshney

**Culture of Animal Cells:
A Manual of Basic Technique,
4th Edition**

2000
ISBN 0-471-34889-9

R. Ian Freshney, Mary G. Freshney
(Eds.)

**Culture of Epithelial Cells,
2nd Edition**

2002
ISBN 0-471-40121-8

Kay C. Dee, David A. Puleo, Rena Bizios

**An Introduction to Tissue-
Biomaterial Interactions**

2002
ISBN 0-471-25394-4

Rolf D. Schmid, Ruth Hammelehle

**Pocket Guide to Biotechnology
and Genetic Engineering**

2003
ISBN 3-527-30895-4

Michael Hoppert

**Microscopic Techniques
in Biotechnology**

2003
ISBN 3-527-30198-4

Oliver Kayser, Rainer H. Müller (Eds.)

**Pharmaceutical Biotechnology:
Drug Discovery and Clinical
Applications**

2004
ISBN 3-527-30554-8

Tissue Engineering

Essentials for Daily Laboratory Work

W. W. Minuth, R. Strehl, K. Schumacher

WILEY-VCH

WILEY-VCH Verlag GmbH & Co. KGaA

Authors

Dr. Will W. Minuth, PhD
Raimund Strehl, PhD
Karl Schumacher, M.D.

University of Regensburg
Department of Molecular and Cellular Anatomy
University Street 31
93053 Regensburg
Germany

Translated by

Renate FitzRoy (Chapters 6, 8)
26 Cairnhill Gardens
St Andrews, Scotland KY16 8BX

Nicole Heath (Chapters 1, 2, 3, 4, 7, 9, 10, Glossary)
Fischergasse 6
69117 Heidelberg

Matthias Herbst (Chapter 5)
Markgräflerstr. 12
69126 Heidelberg

Important Note:

As research and clinical work are constantly expanding our knowledge, we would like to emphasize that when this book was written, all dosage and application specifications reflected the state of the art. However, users are strongly advised to check the instructions that come with the preparations and medical products used and use their own judgement on dosage according to specific recommendations in their own countries.

Finding Literature Using Search Criteria

Due to the rapid growth of information, constantly transforming our knowledge in the area of cell biology and tissue engineering , it made sense to collate a number of search criteria rather than a bibliography. Put in a medical or biological database such as PubMed or Biological Abstracts, these search criteria will lead you to the most up-to-date literature on the subject.

Library of Congress Card No.: Applied for.

British Library Cataloguing-in-Publication Data:
A catalogue record for this book is available from the British Library.

**Bibliografische Information
Der Deutschen Bibliothek**
Die Deutsche Bibliothek lists this publication in the Deutsche Nationalbibliografie; detailed bibliographic data is available in the internet at http://dnb.ddb.de.

© 2005 Wiley-VCH Verlag GmbH & Co. KGaA, Weinheim

Printed in the Federal Republic of Germany.
Printed on acid-free paper.

Composition Mitterweger & Partner, Plankstadt
Printing betz-druck GmbH, Darmstadt
Bookbinding Großbuchbinderei J. Schäffer GmbH & Co. KG, Grünstadt

ISBN-13: 978-3-527-31186-6
ISBN-10: 3-527-31186-6

Preface

Why this book at this time? A number of things have come together. In restructuring our lab, we needed to clear out, organize and archive. A lot of interesting material from the past was lying around that, for various reasons, was not being further investigated and thus had never been published. On inspection of the data and images, we realized that we had actually learned much more from unsuccessful experiments than from the successful ones that had seamlessly fit into the experimental design. When we came upon difficulties, we did not give up. We continually asked new questions and carried out further experiments until we came to logical explanations.

In addition, we have offered many courses in cell and tissue culture as well as tissue engineering over the years, for participants from both Germany and abroad. The participants often asked interesting and fundamental questions which were insufficiently or completely unanswered by previous books. To solve this problem, it was necessary to do a great deal of research in the various databanks. We have sketched, structured and worked the answers to those questions into the text as fundamental information.

Although we train students daily in microscopic anatomy, it has become increasingly evident to us how little is known about the development of functional tissues. However, it is exactly this aspect that is of particular importance for the future production of tissue constructs, from adult cells or stem cells, for use in patients. Socially interactive cell networks must be produced out of individual cells and implanted into the patient as functional tissue, and no health risks should be added in the process.

This book introduces theoretically fundamental and experimental concepts, which should open the door into the field of tissue engineering. Additionally, it should give students, technicians and young scientists a look into the fascinating world of differentiable cells and tissues. We must make clear that we stand at the beginning of a very exciting and future-oriented scientific development. For this reason, we must adjust ourselves to learning about the development of tissues. After sufficient experimentation, and in the course of this decade, tissue engineering will change from a purely empirical to an analytically reproducible science. We will get an overview of each step in tissue development and learn to simulate it experimentally. Apart from molecular biological processes, epigenetic factors and microenvironments will also play a major

Tissue Engineering. Essentials for Daily Laboratory Work W. W. Minuth, R. Strehl, K. Schumacher
Copyright © 2005 WILEY-VCH Verlag GmbH & Co. KGaA, Weinheim
ISBN: 3-527-31186-6

roll. In addition, we must adjust to the fact that it will not be possible to generate functional tissues with cell culture methods.

Will W. Minuth, R. Strehl, K. Schumacher   Regensburg, February 2003

Contents

Tissue Engineering. Essentials for Daily Laboratory Work W. W. Minuth, R. Strehl, K. Schumacher
Copyright © 2005 WILEY-VCH Verlag GmbH & Co. KGaA, Weinheim
ISBN: 3-527-31186-6

1
Developmental processes

Cell, tissue and organ cultures today are no longer to be ignored, for a variety of reasons. For one, in recent years enormous progress has been made in the clarification of molecular and cell biological processes with the help of cultivated cells. Another reason is that without various cell cultures, the industrialized production of many medications and antibodies would be unimaginable. Finally, cultivated cells are repeatedly brought up in discussion as an alternative to animal experimentation.

All the cells of an organism can be isolated from tissue using the modern methods at our disposal today. In addition, nearly all cells can be cultivated without major difficulties for various purposes, both in analytically small as well as technically large scales. The scale can vary from single cells in a droplet to bioreactors with thousands of liters of culture medium. Through these techniques one can build on about 50 years of experimental experience in cell culture. Key phrases for the industrial use and the work associated with it are "cell culture engineering", "metabolic engineering", "bioprocessing", "genomics", "viral vaccines", "industrial cell culture", "medium design", "viral vector production", "cell line development", "process control" and "industrial cell processing". However, almost all of these terms involve a particular type of culture. The cells in question should divide as fast as possible in order to more efficiently synthesize a bioproduct, medication or vaccine. A wide variety of innovative instruments have been developed in recent years for all these techniques. In addition, these methods have been so well optimized that little increase in efficiency can be expected in the next few years. A great deal of information on this topic is available in previously published books.

Tissue culture, and therefore tissue engineering, must been seen very differently. The purpose here is to achieve, or produce, functional tissue and sections of organs through cultured cells. These constructs should support regeneration as implants or be used as bioartificial modules at the patient. Tissue engineering is a relatively young technique, building on 10–15 years of understanding in the field. For this purpose, whole branches of science in the areas of biomaterial research, engineering science, cell biology, biomedicine and individual disciplines in surgery must work closely together.

Considerable progress has been made in the production of artificial tissue with the presently available methods. Nevertheless, it is a fact that the constructs currently produced still do not have sufficient tissue specificity. Liver parenchyma in bioartifi-

Tissue Engineering. Essentials for Daily Laboratory Work W. W. Minuth, R. Strehl, K. Schumacher
Copyright © 2005 WILEY-VCH Verlag GmbH & Co. KGaA, Weinheim
ISBN: 3-527-31186-6

cial modules shows only a fraction of the original detoxification capacity, implanted pancreas cells lose their ability to synthesize insulin over time, kidney epithelia tend not to maintain the necessary barrier and transport functions, and cartilage constructs build an extracellular matrix (ECM) with too little resistance to mechanical load. In addition, proteins that are not typical to the specific tissue are often synthesized by the constructs and can cause inflammation or even a rejection reaction.

In the media, one has the impression that many currently incurable diseases will very soon be treatable with cell therapy, tissue engineering or the manufacturing of organs. It is envisioned is that stem cells will primarily be used. In the spotlight, in particular, are embryonic stem cells whose future significance in this area is still undetermined and whose cell biological capacity seems to inspire enthusiasm without critique. On closer consideration, however, it becomes clear that most current knowledge has been obtained from pluripotent stem cells of the hematopoetic system. Far less experience has been gained in embryonic stem cells from experimental animals and there is very little truly validated experimental data for embryonic stem cells in humans. The existing results in this area often do not seem to be thrilling and highlight many unsolved problems.

There is, also, comparatively little knowledge about the development of totipotent human stem cells. In this case, international research will only show in the coming decade if the promises of many biotechnology firms hold up to critical analysis. The regeneration of functional tissue cannot be solved with isolated stem cells alone. Stem cells, as with all other cells, must first divide in sufficient quantities, form social networks and then develop into specialized tissues, through mechanisms still unknown at this time. These processes are carried out automatically in a developing organism. When trying to simulate these processes *in vitro*, however, one realizes that the characteristics developed in the constructs using the currently available strategies are insufficient.

From our perspective, future key issues to be clarified in tissue engineering are how functional tissue can be generated in culture and how the development of tissue properties can be individually controlled. Artificial tissues will only then be considered a meaningful form of therapy, when a disease can be overcome without harm to the patient. In order to do this, a tissue must exhibit the necessary functional characteristics as a regenerational tissue, an implant or biomodule.

Every day we are faced with all types of functional tissues in the adult organism, in terms of both macroscopic and microscopic anatomy. The adult organism and, therefore, the endpoint of development are fairly well known to us. There are, also, numerous verified discoveries in the early development of humans, as much research has been done in the field of embryonic development and germ tissues. The point and location that a tissue or organ originates from has been specifically studied. Surprisingly little is known, in contrast, about the mechanisms in the development of functional tissues. Understanding this development, however, is key to the production of optimal artificial tissue.

The only available databanks are not very productive if data regarding functional tissue development is requested. It may be surprising, but we could also not find any book about the processes of tissue development. Recently, however, increased activity in this area can be observed. There are various attempts to explain the development of ground tissue with its functional facets through molecular biology. The driving force for this is certainly stem cells. It has been shown that individual functional tissues cannot even be developed out of this type of cell. Only precise understanding of the specific developmental physiology can lead to the generation of tissue.

In the area of regenerative medicine, there are many fascinating and unanswered questions, such as why certain cells in an organism cease to divide after months, years or life-long, whereas other cells are renewed after days. Often these processes even happen side by side in an individual tissue. This alone cannot be explained by the effects of growth factors or morphogenic substances. The microenvironment and cell interactions must have much more effect on the individual regenerational behavior. This means that future perspectives into the developmental needs of tissues must be sharpened and expanded accordingly.

[Search criteria: cell culture organ culture tissue culture tissue engineering]

2
Cells and Tissue

2.1
The Cell

Natural tissue as well as artificial tissue is composed of many different cellular elements and their associated ECM. The cells build multicellular networks and interact with the ECM. Before one can consider the production of artificial tissue, it is necessary to have a fundamental comprehension of cells and natural tissue. The following, however, can understandably convey only certain important aspects of microscopic anatomy.

2.1.1
The Cell as a Functional Unit

Human cells should first be schematically introduced as the smallest functional unit of life. It is generally accepted that a typical characteristic of a living cell is its adequate response to stimuli, such as hormones. Another typical property of cells is that they double their number at regular intervals. This is true for all embryonic cells, as well as all cells of the maturing organism. For cells in a tissue of the adult organism, on the other hand, there are specific differences. Cells in the intestinal epithelium are renewed within a few days, whereas parenchymal cells of the liver or kidney divide only after years. Heart muscle cells and neuronal cells will not normally divide again, even after a lifetime.

The human body possesses around 1×10^{13} tissue cells, living in close contact. In addition, 3×10^{13} blood cells, for the most part in isolated form, can be found in the bloodstream. At the same time, cell size varies widely. The diameter of glia cells (neuronal tissue) is 5 μm, that of sperm cells 3–5 μm, that of liver cells 30–50 μm and that of a human oocyte 100–120 μm.

As with the size, the shape of cells is quite variable. Between the round or spindle shaped and the strict geometric shape of cells in epithelia, all transition shapes can be found. The cell surface can be smooth or uneven. Furthermore, individual surface enlargements from single microvilli to specialized brush borders can be developed. Animal and human cells are surrounded by a selectively permeable membrane (Fig. 2.1), inside which the cytoplasm with the nucleus and other essential organelles

Tissue Engineering. Essentials for Daily Laboratory Work W. W. Minuth, R. Strehl, K. Schumacher
Copyright © 2005 WILEY-VCH Verlag GmbH & Co. KGaA, Weinheim
ISBN: 3-527-31186-6

Fig. 2.1: Illustration of a cell with its organelles: nucleus (1), plasma membrane (2), ER (3), Golgi apparatus (4), mitochondria (5), secretory granules (6), microvilli (7) and centrioles (8).

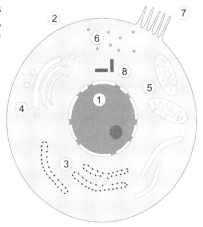

are located. Under light microscopy, the predominantly basophilic cells can easily be differentiated from the mostly acidic cytoplasm.

2.1.2
Plasma Membrane

The plasma membrane is a biological membrane that divides physical/chemical compartments from each other. It is composed of a phospholipid bilayer, through which unipolar molecules such as O_2 and CO_2 can freely diffuse. It serves as a barrier to electrolytes, amino acids and sugars. Under light microscopy, it appears as a trilaminar structure: light–dark–light. Built into this lipid bilayer are numerous proteins that, among other tasks, through targeted transport or as hormone receptors have a mediator function for the information exchange between the cytoplasm and the extracellular environment. A cell membrane, however, is not a mechanically fixed and therefore rigid structure, but rather a fluid, viscous and, accordingly, fragile mantle. The individual phospholipids and the membrane proteins, both, are more or less mobile within this layer. Apart from phospholipids, other lipid molecules, such as cholesterol, are present which provide a certain amount of stability in the bilayer. The outer lipid layer of the plasma membrane contains many glycolipids and glycoproteins, whose sugar residues, oriented outwardly, form their own layer, referred to as the glycocalyx. The proteins that are built into the plasma membrane are made of integral and associated membrane proteins, each with hydrophobic and hydrophilic sections. The hydrophobic sections provide anchoring in the lipid layer, whereas the hydrophilic sections reach out to the extracellular space or into the cytoplasm. Many of these proteins are actually glycoproteins. Functionally, they are transport proteins for electrolytes and amino acids, receptor proteins for hormones or anchoring proteins.

One of the main functions of the plasma membrane is as a diffusion barrier. It can control which molecules pass into or out of the cell by means of various active or passive transport processes. A further function of the plasma membrane specific to tissue cells is its communication ability. Cells are able to communicate with each other over the plasma membrane and build mechanical cell contacts through tight junctions or communication channels through gap junctions. This serves to control cellular exchange, as well as cell recognition or signal processing. These functions are particularly important when social networks develop from isolated cells and, from there, form functional tissue.

2.1.3
Nucleus

With the exception of red blood cells or erythrocytes, all human cells contain a nucleus. The most important component of the nucleus is the chromosomes. They contain the complete set of genetic information. In addition, the nucleus is the control organ for many cell functions. The nucleus, with individual chromosomes, can only clearly be seen under light microscopy during interphase, i.e. between mitotic cycles. In a similar manner, the nucleolus is only observed during this phase. A cell has, as a rule, only one nucleus. However, some cells of particular tissues may have two or even more nuclei. These can be found in the parenchymal cells of the liver, in osteoclasts and in striated musculature.

2.1.4
Mitochondria

The mitochondrion represents the power station of the cell and is a carrier of enzymes, which enable it to produce energy, in the form of adenosine triphosphate (ATP). The characteristic reaction processes in the mitochondria are the energy-producing citric acid cycle and the β-oxidation of the fatty acids. In places where many mitochondria are found within a cell, it can be assumed that synthesis or working processes with increased energy requirements are also taking place there. This process can be identified by, among other things, the fact that the plasma membrane is strongly folded (Fig. 2.2). Within the folds are many mitochondria. Physiological transport investigations in such cells have shown that here increased energy-consuming transport pumps are also inserted, which manage the increased cellular exchange. Such processes can be clearly observed as a morphological correlate on the cells in the salivary glands.

2.1.5
Endoplasmic Reticulum (ER)

The ER plays the decisive role in protein synthesis. Cytoplasmic proteins are built on free ribosomes (polyribosomes), whereas proteins of the plasma membrane as well as secretory proteins are built in the ER. In the cytoplasm, ribosomes can exist individually or in chains, referred to as polysomes. Polyribosomes are connected by a single-strand messenger RNA (mRNA). The oxygen-binding protein, hemoglobin, for

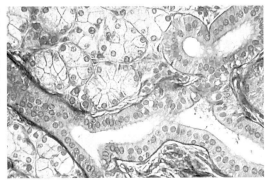

Fig. 2.2: Histological representation of an exocrine gland with part of the duct. The basolateral plasma membrane is largely unfolded. Mitochondria that supply the necessary energy for the pumps in that area are inserted in the folds of the plasma membrane. Due to the folding, the nuclei are crowded into the luminal cell side.

example, is formed on such polyribosomes. Ribosomes involved in the formation of glycoproteins and lipoproteins, on the other hand, do not simply release their protein product into the cytoplasm, but pass it on into the lumen of the ER. The ER is a net-like membrane system of tubules and cisterns, found throughout the cell. It is partly covered with numerous ribosomes and designated as the rough ER (rER). Ribosomes are macromolecules composed of proteins and ribonucleic acids, and not contained within a membrane.

2.1.6
Golgi Apparatus

The Golgi apparatus is found in direct vicinity to the ER. Depending on the cell type, it consists of a varying numbers of dictyosomes and Golgi vesicles. The dictyosomes, or Golgi fields, appear in electron micrographs as stacks of membranous sacs, surrounded by numerous vesicles. In the Golgi apparatus, transport vesicles coming from the ER and containing newly synthesized proteins are processed. As an example, proteins delivered to the Golgi are modified with particular sugar molecules (glycosylation). The end result is glycoproteins or proteoglycans. Frequently, proteins are only biologically active after this step.

2.1.7
Endosomes, Lysosomes and Peroxisomes

Endosomes and lysosomes are a heterogeneous group of organelles, which serve very diverse metabolic processes. Lysosomes are membrane vesicles with particular enzymatic equipment for intracellular metabolic processing, separation and digestion. The metabolic products produced in the lysosomes can be passed on into the surrounding cytoplasm or reused if necessary. On the other hand, lysosomes also serve as a storage place for metabolites that cannot be further broken down. They are then referred to as residual bodies and can be seen as pigment or lipofuscin granula for diagnostic purposes. If the contents of the lysosome enter the cytoplasm uncontrolled, the entire cell, as well as the adjoining cells, can be destroyed by autolysis.

Peroxisomes do not occur in all cells. On the other hand, some cells, e.g. liver cells or tubule cells of the kidney, are particularly rich in peroxisomes. The most important function of these organelles is to house hydrogen peroxide-producing oxidases and catalases, which play an important roll in gluconeogenesis, fat metabolism and various detoxification reactions.

2.1.8
Cytoskeleton

The cytoskeleton (Fig. 2.3) forms the scaffold for other important components of the cell. It consists of microtubules, microfilaments and intermediate filaments. These form a micro-network and function as the skeleton of the cell. Important proteins of this network are tubulin, actin filaments, myosin filaments, the many different keratins, nexins, vimentin, desmin and neurofilaments. Microtubules serve the directed transport of molecules within the cell. Neurons, for example, can possess axons which are 1 m long. Even the synapse, as the end of the neuron, must be controlled by the neural cell body. With the microtubule system, a transport speed of up to 400 mm/day is ensured so that even the most distant end of the cell is provided for.

Microfilaments such as actin filaments and myosin filaments are found in cells in differing quantities. Cells that form extensions, and change form and location exhibit a particularly large amount of microfilaments. Intermediate filaments, such as cytokeratins, build the skeletal system in epithelial cells, and give them their specific shape and stability.

2.1.9
ECM

Most cells produce not only their own organelles, but also proteins of the surrounding ECM. This is an interactive scaffold that provides mechanical stability and cell anchorage, and is also able to control cell functions. In building the ECM, cells synthesize mainly high-molecular-weight fibrous proteins, which are secreted out of the cell and

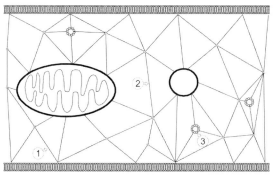

Fig. 2.3: The cytoskeleton of a cell consists of microfilaments (1), intermediate filaments (2) and microtubules (3). From the three-dimensional linkage of these structures, a meshwork results into which the individual cell organelles, such as the mitochondria, are built. Thus, in homogenous tissue cells, the organelles are always found in the same position.

built up in the surrounding environment to form an insoluble network. In epithelia or muscle cells this is a leaf-like basement membrane, whereas connective tissue cells form a three-dimensional network, called the pericellular or ECM. The basement membrane and the pericellular matrix consist mostly of the same protein families; however, due to the varied amino acid sequences, the individual components are differently interconnected. Components of the ECM include the various collagens, laminin, fibronectin and individual proteoglycans. In many tissues the ECM is soft and elastic, whereas mechanically strong structures are formed in tendon, cartilage and bone.

2.1.10
Cell Cycle

Cells must proliferate in order for tissues to develop, as well as for the replacement of dead cells through regeneration in the adult organism,. This is carried out within the framework of the cell division cycle (Fig. 2.4). First, cells double their contents and replicate their DNA in interphase. Next, the cells divide in mitosis. A cell in interphase can usually be recognized by the clearly defined nucleolus. If the decision is reached for a cell to divide, the cell continues into the G_1 phase, where the formation of important molecules, such as RNA, proteins and lipids, take place within about 24 h. In addition, the volume of the cell increases. In the subsequent S phase, the DNA in the cell is replicated. If this important phase is complete, the cell continues into the G_2 phase. Replication of the DNA is completed and everything is prepared for the actual division of the cell.

Mitosis itself takes about 4 h. In prophase, DNA/histone complexes condense into 46 chromosomes. The mitosis spindle is formed on the developing centrioles. The nuclear envelope and the nucleolus dissolve. Phosphorylation of the lamina in the nuclear membrane follows and, eventually, reusable vesicles are formed again. In metaphase, chromosomes arrange themselves in the equatorial plane or at the site of future division, each chromosome consisting of two sister chromatids. At this stage, long and short sections of the individual chromosomes are clearly visible under light microscopy. As the process continues, the chromosomes divide into the sister chromatids and, with the assistance of motor proteins, are transported along the mi-

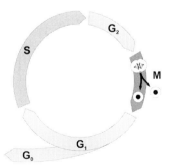

Fig. 2.4: Schematic of the cell cycle, which is divided into the G_0, G_1, S and G_2 phases. The actual division of the cell takes place in the M phase.

crotubules to the centrioles during anaphase. In the following telophase, a new nuclear envelope is synthesized. Cell division is terminated by the production of a ring of actin and myosin filaments, which cleaves the cell in two. In this phase of cytokinesis, each daughter cell receives one of the newly synthesized nuclei and half of the cytoplasm, along with necessary organelles.

Depending on the tissue type, cells can divide within days or only after months or years. In addition, some cells will not divide again during the life of the organism. Non-proliferating cells are said to be in G_0 phase.

[Search criteria: cell cycle mitosis division interphase]

2.2
Tissue Types

The development of cellular networks in complex organs is reflected in the structural and functional characteristics of tissues. Tissue is not only an accumulation of individual cells, but consists of defined cellular and specific extracellular structures. Both parts are functionally irreplaceable.

Surprisingly, humans only possess four different kinds of basic tissue – epithelia, connective tissue, muscle tissue and nervous tissue. From these come four completely different functions, such as the division of the organism from other compartments, the connection of structures, movement and control.

No organ of the body consists of only one basic tissue. Nearly all need each of the four tissues in a particular arrangement in order for each special function to become effective. The vascular system is one example. It consists of epithelial tissue, which lines the vessel lumen, smooth muscle tissue, in order to change the blood flow, nervous tissue, for controlling the rate of blood flow, and connective tissue, which connects the individual structures to each other and the surrounding environment. Tissues can consist of either homogenous or quite different cell types. Particularly characteristic is that clearly defined social contact (sometimes close, sometimes loose) is cultivated for the maintenance of specific functions in individual tissues.

Typically, one finds many mobile cells in tissue, such as leukocytes, plasma cells and macrophages, which react to cell metabolites, antigens or bacterial infection, and thus serve in immunological defense. Accordingly, few of these cells are to be observed in healthy tissues, whereas the cell number drastically increases during illness.

[Search criteria: tissue muscle epithelium connective neural]

2.2.1
Epithelia

The epithelia consist of geometric, spatially closely connected cells, which are anchored to a basement membrane (Fig. 2.5). Virtually no intercellular substance is to be found between epithelial cells.

Fig. 2.5: Structural drawing of single squamous and pseudostratified epithelia. It is typical of all epithelia that the cells have a particularly close relationship to neighboring cells and are anchored to a basement membrane. (A) Schematic illustration of cuboidal epithelia. The basal side of each cell is anchored to the basement membrane and the apical plasma membrane borders the lumen. The lateral cell borders are in contact with the adjoining cells. (B) With the pseudostratified epithelia, several cell types are present. All cells are anchored to the basement membrane, but not all reach the lumen, giving the illusion of multiple layers. (C) With the stratified epithelia, only the basal cell layer has contact to the basement membrane.

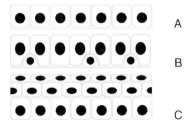

Epithelial tissue forms a multitude of biological barriers – the central function of epithelia in an organism. It covers surfaces in a layer of closely connected cells, thereby forming a barrier between air- or liquid-filled compartments of the body. For this reason, it is the epithelia alone that determines what is taken up by or excreted by the body at the cellular level. It regulates the uptake of gas and fluid, and output by means of active or passive transport mechanisms. Epithelial cell layers have, for the most part, no gaps between cells and, with the exception of the stria vascularis in the inner ear, have no blood vessels.

Epithelial cells sit with their basolateral side on a basement membrane. The basement membrane is the structural element which separates epithelia from the connective tissue beneath it. If this barrier is no longer functionally intact, carcinoma cells can leave the epithelial compartment and infiltrate the connective tissue.

At their surface, epithelia exhibit various cell differentiations. On the one hand, they may have a more or less smooth surface. On the other hand, there may be a dense brush border for surface area enlargement or kinocilia for increased transport function. Characteristic of all epithelia is their polarization. This means that each cell has one side oriented toward the lumen and one toward the basement membrane, through which uptake and output of molecules take place.

2.2.1.1 Building Plans of Epithelia

Epithelia lining the body surfaces may consist of one layer (simple), multiple layers (stratified) or one layer appearing to be multiple layers (pseudostratified). Epithelia can take on completely different forms. They may be flat (squamous), cuboidal or cylindrical (columnar). Simple epithelia are characterized by the contact of all cells with the basement membrane. Flat epithelia form the typical squamous-shaped epithelia which occur in the lining of blood vessels as endothelial cells. Vascular endothelial cells are continually exposed to the bloodstream and therefore need to be anchored particularly firmly to the basement membrane (Fig. 2.6). Adhesion molecules for leukocytes, in the form of selectins, are located on the endothelial surface are. These are able to bind sugar molecules on leukocytes and thus facilitate their exit from the bloodstream. Endothelial cells further possess contractile filaments, with which they can regulate the width of their intercellular gap, to a certain extent. Endothelial cells additionally

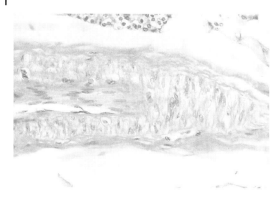

Fig. 2.6: Microscopic view of an arteriole – diagonal cut. The lumen is left in the center, lined by endothelium. Numerous smooth muscle cells are found in the media of the vessel wall.

produce nitrogen oxide (NO), which leads to a reduction in the smooth muscle tone surrounding blood vessels and thus to a increase in flow.

Single squamous epithelia, however, occur outside of blood vessels as well. They line the alveolar space in the lung, and provide a short diffusion gap for carbon dioxide and oxygen, due to their flat shape. Beyond that, this type of epithelia is found both in the thin part of the loop of Henle (kidney) as well as in the epithelial lining of the serous pleura and the peritoneum.

Cuboidal epithelia can be recognized under light microscopy by the fact that their cell width and height are about the same (Fig. 2.5A). Cuboidal cells are found, among other places, in renal tubule structures, where they serve in transport processes, and in urine production or in salivary glands, where they are active in saliva production. Likewise, cuboidal cells in the follicles of the thyroid are shown to work as storers and donors of hormones. Columnar epithelia are higher than they are wide, and line the lumen of the whole small and large intestine in the form of enterocytes, for example, where they serve the uptake of nutrients.

In common with simple squamous epithelia, pseudostratified epithelia cells are also in contact with the basement membrane (Fig. 2.5B). However, they differ from simple squamous epithelia in that not all cells reach the upper surface of the epithelia and the cells, as well as their nuclei, are at different levels. The pseudostratified epithelia of the respiratory tract are specialized with moveable kinocilia on their surface and are therefore referred to as ciliated epithelia (Fig. 2.7).

Cells are found in three different layers in pseudostratified epithelia, one above the other (Fig. 2.5C). Basal cells are anchored to the basement membrane and are not in contact with the epithelium surface. From this basal cell layer, epithelial cells are regenerated continually and life-long from stem cells. In the intermediate zone, immediately above the basal cell layer and in the luminally situated stratum superficiale, the cells are no longer in contact with the basement membrane. The cells of the outer epithelium surface are periodically sloughed off. In stratified squamous epithelia, the surface cells are flattened and, in contrast to the basement cells, oriented parallel to the epithelium surface. The basal cells are usually cuboidal to columnar. Cells of the intermediate zone lose this orientation. They become polygonal and their nuclei become more parallel to the epithelium surface. The mucous membranes from the oral

Fig. 2.7: Histological illustration of pseudostratified ciliated epithelium in the respiratory tract. The epithelia borders the airway. The luminal epithelial side with kinocilia serves the function of cleaning and moves dirt particles toward the oral cavity.

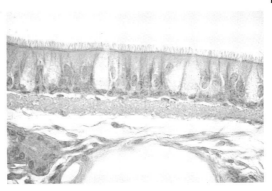

cavity to the lower third of the esophagus (Fig. 2.8), of the vagina, as well as transition areas of the urogenital and digestive tract to the outer skin exhibit this stratified squamous epithelia.

In contrast to the stratified squamous epithelia in the mucous membrane of the mouth, the stratified squamous epithelia of the outer skin is keratinized. The stratified squamous epithelium of the skin shows characteristics of the dynamic keratinization process, which begins in the stratum granulosum. The regenerative basal cells lie in contact with the basement membrane. However, they do not occur here alone, e.g. neighbor melanocytes, which through their pigment are responsible for the brown coloring of the skin. In the intermediate zone is first the stratum spinosum. The spiky appearance of this cell layer is due to the numerous occurrences of desmosomes, into which bundles of condensed cytoskeletal components lead. These cellular characteristics serve as protection against shearing stress which can affect the outer skin. In the next layer, the stratum granulosum, are the cytoplasmic keratohyaline granula, which contain the protein filagerin, visible under light microscopy, to a high degree. Apart from the transverse cross-linking of proteins, the superficial cells experience organelle degradation, including that of the nucleus. In the end, the cells of the stratum corneum consist only of a closely packed keratinized substance, surrounded by a modified cell membrane. The Langerhans cells, which carry out immunological tasks, are also next to the melanocytes.

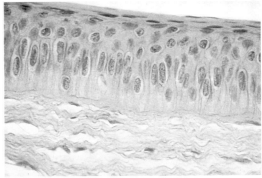

Fig. 2.8: Histological illustration of the non-keratinized stratified epithelium of the oral cavity. The epithelial cells form close cellular networks. Thus, a biological barrier between the luminal and basal sides of the epithelium develops.

A further stratified epithelium is the transition epithelium, also known as the urothelium, which is exposed to urine on its luminal side. Biologically aggressive substances, such as urea and the changing pH of the urine, have led to its particularly pronounced tight junctions and the condensed cytoskeletal elements found on the luminal side of its cells. These consist of actin filaments and intermediate filaments, as well as uroplakin. The superficially located cover cells have a polygonal form and branch-like cell extensions, which should reach the basement membrane. In this sense they differ substantially from the other stratified squamous epithelia. The name transition epithelia suggests that the epithelium can be stretched to accommodate different volumes and because of this can present a large range of cell heights.

[Search criteria: tissue epithelial morphology histology]

2.2.1.2 Glands

Glands result from cells of the surface epithelia budding into the connective tissue beneath. If the gland forms a duct, one speaks of an exocrine gland (Fig. 2.9A and B). If, however, this developed epithelium loses its contact with the original surface epithelium, the island remaining in the connective tissue can secrete only into the

A

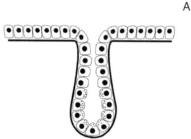

B

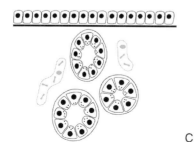

C

Fig. 2.9: Schematic illustration of gland formation. (A) Exocrine and endocrine glands are formed from the simple squamous embryonic epithelium. (B) A glandular duct results from the invagination of the epithelium into the connective tissue beneath it. If the lumen maintains contact to the surface, then an exocrine gland is formed. (C) If the encompassed epithelial cells lose contact to the lumen, an endocrine gland develops. At the same time, capillaries are increasingly developed in this area.

interstitium and into capillaries. One speaks of the development of an endocrine gland and of internal secretion, whereby the secretion contains hormones (Fig. 2.9C).

Glandular tissue consists of epithelial cells, which form a substance and then secrete this from the cell. The basolateral cell side remains in constant close contact with blood vessels, since it needs to take up numerous nutrients from the blood for synthesis. Secretion takes place on the luminal side of the acinar epithelia in the salivary glands, whereas the hormone is always delivered toward the capillary with endocrine glands.

If a connection between the epithelia grown into the connective tissue and the surface epithelium remains, the secretion formed in the gland will exit through a duct. The secretion can then be modified by special cells of the duct in terms of its water and electrolyte composition, similar to in the kidney. This process is possible, for example, in duct epithelia of the parotis (Fig. 2.10). The cells of exocrine glands are polarized, since they take up material from the interstitium over the basolateral side and secrete from it, which is then transferred to the luminal side into the duct. Whole organs, like the salivary glands, can have a purely exocrine function.

Apart from purely exocrine functions, endocrine output can be found in a gland. The classical example is the pancreas with the endocrine islets of Langerhans, which deliver insulin and glucagon into the blood stream in order to regulate sugar metabolism. The exocrine portion of the pancreas produces digestive enzymes, such as amylase and lipase, which are then delivered into the ductus pancreaticus and further into the duodenum.

The end sections of exocrine glands are distinguished by their type of secretion – serous, mucous or mixed seromucous end sections. The epithelia of the gland show corresponding histological characteristics (Fig. 2.10). The cells of the serous end sections possess a round nucleus, which lies in the center of the cell. The cytoplasm presents itself as homogenously reddish in routine staining. The serous secretion is non-viscous and enzyme-rich. In the cells of mucous glands, on the other hand, the nucleus is markedly flattened and lies near the basolateral side of the cell. The cytoplasm appears foamy and whitish. The secretion is more viscous and contains fewer enzymes than the serous secretion. A mucous gland cap sits at the end of some serous end sections and these areas are then designated as Ebner half moons.

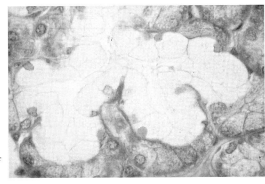

Fig. 2.10: Light microscopy view of a salivary gland, which is composed of mucous and serous acini.

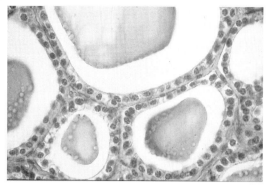

Fig. 2.11: Histological representation of a thyroid. The epithelium forms balloon-shaped follicles, which are filled with colloid in their lumen.

Secretions are delivered by glandular cells in completely different ways. The merocrine form of secretion is based on exocytosis. The intracellular secretory vesicles fuse with the luminal plasma membrane, whereby the secretion is delivered outwardly, without any loss of the cell membrane. In apocrine secretion, the apical portion of the cell, containing secretory products, is pinched off. This process then includes a structural loss to the cell. With holocrine secretion, the secretory product in the cell is released as the cells apoptose and are sloughed off.

The glandular end sections can also take on different forms. In principle, they can be tubular or coiled and tubular. Acinar end sections have a sac-like form. It may be that both forms are found in a gland, with large common end sections. This is then referred to as a compound gland.

Usually the hormone-producing, i.e. endocrine, cells are not polarized. An exception here is in the thyroid, where polarized epithelium is exhibited in an endocrine gland (Fig. 2.11). Here, polarization serves the storage of hormones, which can be mobilized as necessary for delivery into the interstitium.

[Search criteria: glands morphology histology mucous serous seromucous]

2.2.1.3 Epithelia in Sensory Perception

Sensory epithelia are groupings of cells tat can receive and transmit stimuli. In the retina they serve vision; in the inner ear, hearing. They are taste cells (Fig. 2.12), and mechanoreceptors in the outer layer of the skin and on the roof of the nasal cavity are the olfactory epithelium.

In principle, the receptors of sensory epithelia can be divided into primary and secondary groups. Primary sensory cells receive a stimulus on one side of a cell and pass on the excitation over its own axon. This is the case in olfactory cells (Fig. 2.13). They can also be described as nerve cells, which express receptors for certain olfactory molecules at one end. Thus, the olfactory epithelium is the only place in the body in which a nerve cell has direct contact with an exposed surface. Secondary sensory cells, on the other hand, have one sensory end with the appropriate receptors, but are connected synaptically to nerve cells on the other end, such as in the taste epithelium.

Sensory cells never comprise the epithelium alone, but always occur in combination with basal cells and support cells. Basal cells are thought to serve as stem cells for both

Fig. 2.12: Histological cutout from the surface epithelium of the tongue with integrated taste buds.

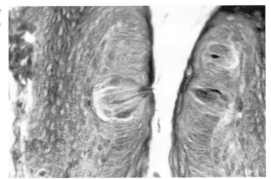

the sensory and support cells. Support cells are needed to build and maintain the necessary environment for the reception. On the other hand, support cells have been described to have a reserve function or have the ability to transdifferentiate, and be converted into sensory cells.

[Search criteria: sensory epithelium morphology histology]

2.2.2
Connective Tissue

Connective tissue occurs all over the human body and in many types. It consists of a variety of cell forms, which partly occur in isolated form, as with bone, and partly aggregate into cell families, as with cartilage. Differently sized areas that can be filled with mechanically loadable intercellular substance or liquid are found between the cells (Fig. 2.14).

As the name suggests, connective tissue connects completely different structures. Frequently, the importance of connective tissue in the individual organs is underestimated and the organ parenchyma, with its functional cells, is placed in the foreground. Just as significant is the stroma of connective tissue in the individual or-

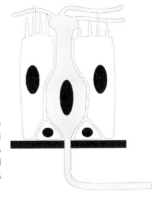

Fig. 2.13: Schematic illustration of the olfactory epithelium on a perforated basement membrane. The sensory cell has long microvilli on the luminal cell side for stimulus perception. The excitation is passed on over an axon on the basolateral side. On the lateral cell sides, the sensory cell is surrounded by columnar supporting cells as well as by basal cells.

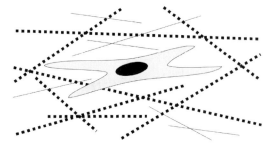

Fig. 2.14: Schematic illustration of a fibroblast, which is surrounded by ECM proteins such as collagen, fibronectin and proteoglycans.

gans. It forms the matrix for organ structure, and brings supply and regulating structures to the parenchyma. Thus, stroma and parenchyma are two irreplaceable components, and only their combination makes the complex functions of an organ.

[Search criteria: connective tissue morphology histology]

2.2.2.1 Variety

Aside from the organ stroma, there is the connective and supporting apparatus of the body – fat, cartilage and bone tissues, with their own specific characteristics. In view of the different kinds of connective tissue, it becomes clear that this tissue has not only a cellular, but also a particular extracellular component. The qualitative and quantitative relationship of cellular to extracellular components can be developed very differently.

In addition, the fixed cells of the connective tissue are differentiated from the free or mobile cells. Connective tissue cells, also, differ in their degree of differentiation and thus in their functional state.

Free connective tissue cells include leukocytes, plasma cells, macrophages and mast cells. These cells leave the bloodstream and settle in the connective tissue to varying degrees, whereby a completely different distribution in the connective tissue results. An intensified concentration of leukocytes, plasma cells and macrophages in connective tissue can be observed during an inflammatory reaction.

The individual connective tissue cells develop out of mesenchymal cells (Fig. 2.15). The immature cells are referred to as fibroblasts, chondroblasts or osteoblasts. They are primarily concerned with constructing the ECM.

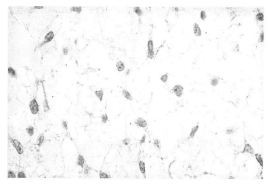

Fig. 2.15: Histological representation of mesenchymal cells, which are surrounded by a fiber-poor ECM. All types of connective tissue can develop from these.

Fibrocytes, chondrocytes and osteocytes, on the other hand, are found in mature tissue. The fibroblast, for example, is involved in the synthesis of collagen in tendons, ligaments, menisci and joint capsules. Under these circumstances, the fibroblast has a large, oval nucleus with a defined nucleolus. The cell borders have many projections and the ER is distinctly noticeable in the cytoplasm. The fibrocyte, on the other hand, is less concerned with synthesis. It controls and supervises the finished ECM in the mature tissue. A fibrocyte is recognizable by its spindly form and narrow, oblong nucleus. While fibroblasts divide frequently, fibrocytes are found in the postmitotic stage. During injury to the connective tissue, fibrocytes can, to a limited degree, differentiate into fibroblasts in response to altered environmental conditions.

A fixed connective tissue cell creates the ECM, which consists of structured and unstructured parts. The structured part consists of fibrous material, while the unstructured section can be described as an amorphous ground substance. Apart from proteolysis and glycoproteins, the amorphous ground substance also contains interstitial fluid. The composition of the interstitial fluid corresponds to the composition of electrolytes and other soluble substances of the blood plasma. Under pathophysiological conditions, the volume of interstitial fluid can rise substantially. All substances that are exchanged between cells and the bloodstream must use the interstitial fluid as a transport medium. The glycoproteins and proteoglycans, which appear amorphous under light microscopy, provide mechanical stability to the tissue. They have a large capacity for binding water, through which cartilage gets its elastic quality.

The fibers of the ECM are divided into collagen, reticular and elastic types, built by the individual connective tissue cells. The production of collagen fibrils takes place in intracellular and extracellular processes. Intracellularly, the synthesis of different polypeptide chains takes place, which through twisting into a triple helix lead to the production of pro-collagen. The pro-collagen triple helix is released by exocytosis. Register peptides are split off from pro-collagen extracellularly, whereby the resulting tropocollagens aggregate into microfibrils and, finally, into collagen fibers, with particularly tensile characteristics. A typical example of this is in the tendon (Fig. 2.16).

There are about 25 different types of collagen. Most connective tissue forms consist not only of fibers from an individual collagen type, but rather from various types,

Fig. 2.16: Histological longitudinal section of a tendon. The darkly stained spindle-shaped nuclei of the fibrocytes can be identified. Between them are arranged parallel bundles of collagen type I fibers

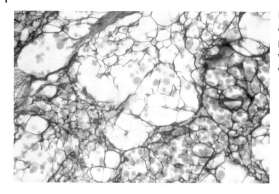

Fig. 2.17: Silver impregnation of reticular fibers in a lymph node. Microcompartments form in the fiber gaps, in which the lymphocytes, along with other material, settle.

although one type commonly prevails. Loose connective tissue contains, as an example, individual branched fibers of the type I collagen and forms the stroma of organs. Tendons, on the other hand, are densely fibrous (Fig. 2.16). The individual fibrocytes lie pushed between the fibers, which leads to a wing-like shape, and are referred to as tendon cells. Collagen II is found in hyaline cartilage. It is important for the microstructuring that, in combination with proteoglycans, gives cartilage its elasticity. Collagen IV is exclusively found in the basement membranes of epithelia, where it provides a place for cells to adhere. Reticular fibers are made from collagen III. They are characterized by the fact that they can be contoured with silver salts. They are, therefore, also called argyrophilic fibers. Reticular connective tissue forms the matrix of many lymphatic organs, like the spleen, the lamina propria of the intestine and the lymph nodes (Fig. 2.17). In this special matrix, the lymphatic cells are held at a distance from one another, allowing their entire surface to be moistened by interstitial fluid. At the same time, this principle prevents the cells from being injured by strong compression.

Elastic fibers are not made up of collagen molecules, but are mainly formed of elastin and fibrillin. The coiled elastin molecules give the elastic fibers their flexibility. Elastic fibers are primarily found in arteries near the heart and lung alveoli.

[Search criteria: connective tissue collagen elastic reticular fibers]

2.2.2.2 Fat Tissue as Storage

Adipose tissue represents a special form of connective tissue (Fig. 2.18). It is common to fat cells and fibrocytes that both develop from the same mesenchymal progenitors. The matured fat cells (adipocytes) are found in two forms in the human body – as the univacuolar, white fatty tissue occurring in many areas of the body (Fig. 2.18A) and as the multivacuolar, brown fatty tissue occurring mainly during infancy (Fig. 2.18B).

One can easily imagine that the univacuolar fat cells are very fragile. As components of the structural and storage fat, however, it is necessary for them to withstand a substantial mechanical load. A meshwork of reticular fibers around each fat cell, therefore, acts to stabilize the cell (Fig. 2.19).

Fatty or adipose tissue fulfills very different functions in the human body. It represents the largest energy reserve in the body, aside from glycogen in the liver and ske-

Fig. 2.18: Schematic illustration of a univa-cuolar fat cell (A) and a multivacuolar fat cell (B). The fat reserve is in the center of the uni-vacuolar cell. Thus, all of the cytoplasmic orga-nelles are shifted to the periphery of the cell.

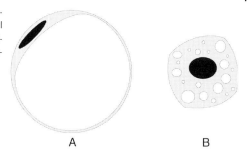

A   B

letal muscle, whereby fats are stored in the form of triacylglycerols. Furthermore, adipose tissue determines the shape of the body with niche and structural fat. On the soles and palms, it functions as structural fat with a mechanical padding function. Since adipose is a bad heat conductor, subcutaneous adipose tissue is also insulating.

The univacuolar fat cell possesses a large fat droplet in the cytoplasm, which pushes the cell organelles completely to the side (Fig. 2.18A). The nucleus is thereby flattened. With the use of routine staining, fixation with xylene and alcohol dissolve the fat droplet, leaving an empty-appearing vacuole. The fat vacuole along with the flattened nucleus is reminiscent of a signet ring. The adipose tissue is also well vascularized, so that fats can be taken up or removed.

Multivacuolar adipose tissue serves, among other functions, in the production of heat during infancy from the storage fats. This adipose tissue is particularly well vascularized and through the cytochrome is found in the many mitochondria (brownish colored). The fat cells of this brown adipose tissue exhibit multiple, small fat droplets in the cytoplasm and the round nucleus sits centrally (Fig. 2.18B).

[Search criteria: adipose tissue histology morphology fat]

2.2.2.3   Bone and Cartilage as Support Tissue

The physical support of the body is mostly the function of cartilage and bone. Cartilage is a bradytrophic tissue, which is not innervated or vascularized (Fig. 2.20). Its nutrition comes through diffusion from the surrounding tissues. The cartilaginous tissue is, in some places, surrounded by a perichondrium, consisting of collagen fibers and

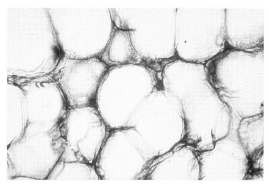

Fig. 2.19: Silver impregnation of the three-dimensional reticular meshwork that surrounds each individual fat cell.

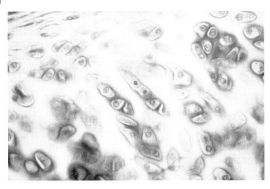

Fig. 2.20: Hyaline cartilage during development – microscopic view. The discrete distance between neighboring chondrones is identifiable. A mechanically loadable ECM develops in the interterritorial areas.

mesenchymal cells. Cartilage consists of chondrocytes, which are walled in by a specialized ECM. Up to 10 cartilage cells, stemming from a chondrocyte, can be found in a lacuna, referred to then as an isogenous group (Fig. 2.21). Chondrocytes are roundish and have a well-developed synthesis apparatus, through which the production of collagen II, proteoglycans, hyaluronic acid and chondronectin takes place. These components and their microstructuring lead to a high level of mechanical stability within the ECM, with a water-binding capacity adapted to changing physiological conditions.

Directly surrounding the chondrocyte is a capsule of cartilage (type VI collagen) and an outwardly bordering, special area of the ECM. Together with the chondrocytes, it is referred to as the territory. Everything outside of the territory is referred to as extra-territorial matrix. Depending on the number of cartilage cells and the composition of the ECM, three types of cartilage can be identified (Fig. 2.22).

Hyaline cartilage occurs in many places of the body, as it forms the joint cartilage and is able to withstand a large amount of mechanical stress. Furthermore, it plays a crucial role in skeletal development, since nearly the entire growth matrix for the skeleton begins with hyaline cartilage and is later replaced by bone. This process is the beginning of replacement bone.

Elastic cartilage is similar in structure to hyaline cartilage. The chondrocyte, however, produces large quantities of elastic fibers for the ECM, from which its deformability results (Fig. 2.23). Elastic cartilage is found in the flexible part of the ear lobe and in the epiglottis of the larynx.

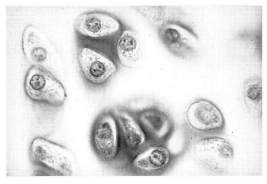

Fig. 2.21: Individual chondrones in hyaline cartilage. The chondrocytes lie together in isogenous groups.

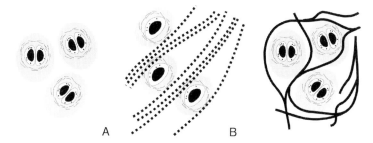

Fig. 2.22: Schematic illustration of chondrones and ECM in hyaline cartilage (A), fibrous cartilage (B) and elastic cartilage (C).

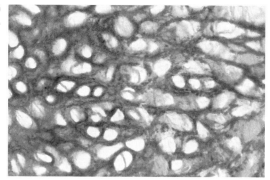

Fig. 2.23: Histological representation of elastic cartilage with numerous fibers.

Type I collagen fibers dominate in the ECM of fibrous cartilage, resulting in compression resistance and tensile strength (Fig. 2.24). With this tissue, only individual chondrocytes are found in the lacunae. Fibrous cartilage is in the pubic symphysis and in the annulus fibrosus of the intervertebral disks. In the latter case, it surrounds the gelatinous nucleus pulposus.

Apart from the different types of cartilage, bone is the mechanically stabilizing element of the movement apparatus. It also it accommodates the cells of the blood-building bone marrow. In addition, it serves as calcium and phosphate storage. During

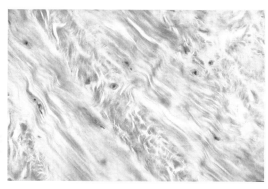

Fig. 2.24: Histological representation of fibrous cartilage. It can be seen that only individual chondrocytes occur in lacunae. It is typical that the interterritorial areas are filled with parallel fibers.

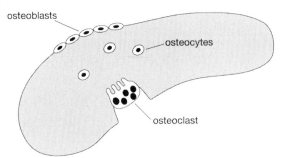

osteoblasts

osteocytes

osteoclast

Fig. 2.25: Schematic illustration of a trabecula, which is developed by osteooblasts and osteocytes. Osteoclasts work antagonistically and break down the hard substance again.

bone development, fibers of bone are formed by osteoblasts (Fig. 2.25). They first synthesize a collagen-containing matrix, which is serves as a scaffold necessary for mineralization. During this process the osteoblasts wall themselves into the mineralized matrix. In so doing, they take on a roundish form and from this point on are referred to as osteocytes. Frequently they remain connected to neighboring cells over thin cytoplasmic extensions.

Bone tissue contains is another type of cell that is responsible for bone resorption – osteoclasts (Fig. 2.26). Their resorption activity is regulated by parathyroid hormone, a hormone produced in the parathyroid gland. Precursor cells of the osteoclasts fuse together, resulting in osteoclasts with up to 60 nuclei. They acidify the surface of the bone through proton pumps, dissolving the hydroxyapatite. This function can also be taken over by some osteolytic osteocytes.

In building lamellar bone, osteocytes become enclosed in lacunae (Fig. 2.27). Radiating from the lacunae are canaliculi, through which fine projections from the osteocytes reach. Through gap junctions they are able to communicate with other cells, including exchanging metabolic products. The matrix situated between osteocytes contains much type I collagen, aside from proteoglycans and glycoproteins. It is believed that particular glycoproteins, such as osteocalcin and sialoprotein, carry out the mineralization of type I collagen fibers. The resulting hydroxyapatite crystals create the hardness of the bone, whereas type I collagen fibers produce its resistance to tensile stress, put on by mechanical challenge to the bone.

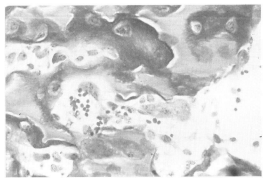

Fig. 2.26: Microscopic view of trabeculae. The many osteooblasts at the edge of the bone matrix can be identified. Osteocytes are enclosed within the matrix. A multinucleated osteoclast is present in the center.

Fig. 2.27: Schematic illustration of an osteon. The Haversian canal can be seen in the center and three lamellas are concentrically located. Between the lamellae are the osteocytes that communicate with one another over very fine cell extensions.

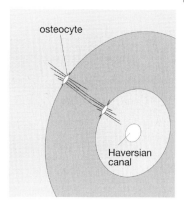

The outside surface of bone is covered by a specialized tissue – periostium. The medullary cavity is lined by the endostium. In the developed lamellar bone, blood vessels penetrate from the outside, through the Volkmann's canals into the bone. There, they connect to the perpendicularly running Haversian canals. The blood vessels in the Haversian canals form the center of an osteon, which serves as the fundamental unit of lamellar bone (Fig. 2.28). Several mineralized lamellas are concentrically arranged around the Haversian canals. Between the lamellas, the osteocytes are in lacunae, connected with one another via canaliculi. The areas between the concentrically formed osteons are filled with general lamellas and connecting lamellas.

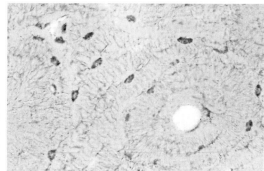

Fig. 2.28: Cross-section preparation of lamellar bone. The Haversian canal with concentric lamellas can be identified. Between them are the dark appearing lacunae, which are settled by osteocytes.

[Search criteria: bone tissue morphology histology]

2.2.3
Muscle Tissue

Contractible cells are introduced in this tissue, which enable movement or the development of tension in skeletal elements, in the heart or in many organs and blood vessels. Histologically, smooth, heart and skeletal musculature are distinguished from one another.

The cellular elements of muscle tissue cause an active contraction. This can only happen in close cooperation with connective tissue, which leads nerve fibers and blood vessels to the muscle cells, and surrounds the muscle tissue. Furthermore, connective tissue forms the connection between the individual muscles and bone in the form of tendons. The different muscle tissue types are relatively simple to differentiate by light microscopy on the basis three criteria (Tab. 2.1):

Tab. 2.1 Diagnostic criteria for the differentiation of skeletal, heart and smooth muscle tissue by light microscopy.

	Striation	Location of nucleus	cappilarization
Skeletal musculature	yes	peripheral	normal
Heart musculature central	yes	central	strong
Smooth musculature	no	central	normal

[Search criteria: muscle tissue histology morphology]

2.2.3.1 **Cell Movement**

Skeletal muscle belongs to the voluntary musculature and can be controlled consciously. Skeletal muscle consists of enormous muscle cells, which can be up to 10 cm long and 0.1 mm wide, and have numerous nuclei (Fig. 2.29). Because of their length, they are referred to as muscle fibers. Skeletal muscle is easy to identify under light microscopy. Striations due to the organized structure of the contractile apparatus can be identified in a longitudinal section, resulting in isotropic and anisotropic areas.

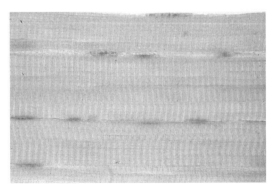

Fig. 2.29: Microscopic representation of skeletal muscle. The muscle fibers, shown in longitudinal section, enable contraction and are recognizable by their striations, among other characteristics, in this case. In order not to impair the contraction, the nuclei lie peripheral to the muscle fiber, displaced by the contraction apparatus.

The main part of the muscle fiber is the myofibrils. For this reason, the cytoplasm, or sarcoplasm in muscle cells, can hardly be seen under light microscopy. Along with the many mitochondria between the myofibrils are the T-tubules. T-tubules are invaginations of the plasma membrane into the sarcoplasm, in contact with the calcium-storing sarcoplasmic reticulum, which surrounds the microfibrils. The sarcoplasmic reticulum is a branched network, developed out of the cisterns of the ER.

In the muscle fibers, which are oriented along the long axis, are the myofibrils. The myofibrils contain sarcomeres, whose borders defined by their Z-disks, contain parallel-oriented actin filaments and myosin filaments. These filaments can slide past one another. By increasingly overlapping, the contractile filaments, actin filaments and myosin filaments shorten the sarcomeres containing myofibrils, which are lined up one after the other. A visible muscle contraction results from the summation of all myofibrils in a muscle.

In controlling a muscle contraction, a membrane depolarization is initiated over the synaptic connection between the muscle fiber and a nerve cell, which is passed over the T-tubules to the inside of the muscle fiber (Fig. 2.30). The functional contact of the T-tubules with the myofibril-surrounding sarcoplasmic reticulum leads to a membrane depolarization through Ca^{2+} release. The Ca^{2+} release causes myosin filaments and actin filaments to form cross-bridges, with a change in position of the myosin heads in relationship to an actin molecule, whereby a shift between actin filaments and myosin filaments takes place, and thus a contraction. The dissociation of actin filaments and myosin filaments takes place only in the presence of ATP.

The synaptic connections between nerve fibers and striated muscle fibers are called motor endplates (Fig. 2.31). A motor unit is defined as all muscle fibers that are con-

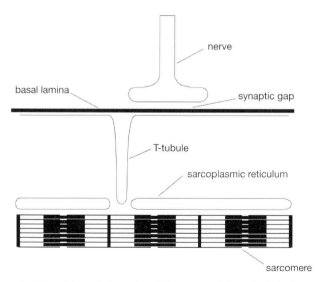

Fig. 2.30: Schematic illustration of the motor end plate of a skeletal muscle fiber. The nerve impulse for contraction arrives over the synaptic gap and the T-tubule system of the sarcomere. Here, the contraction takes place after the impulse is transmitted.

Fig. 2.31: Microscopic view of axons that form a motor end plate and thus a synapse on the surface of muscle fibers.

trolled by the same motor neuron. Motor units with a large number of muscle fibers lead to mass movement. Small motor units are to be found around the outside eye musculature, for example, where the finest changes in movement are possible.

Each muscle fiber is surrounded by a web of reticular fibers, which are anchored to the basement membrane of the muscle fiber. This association of connective tissue around the muscle fiber is known as the endomysium. The endomysium guides vasculature and nerve fibers to the muscle cells. The perimysium, on the other hand, combines multiple muscle fibers into a muscle fiber bundle, called a fascicle. The epimysium lies directly over the entire muscle and is separated from the tough muscle fascia by a gap, which allows for movement.

In the transition from muscle to tendon, collagen fibrils penetrate the numerous invaginations of the muscle fiber plasma membrane, building a particularly strong connection between muscle and tendon. The tendon transmits the muscle contraction to the bone.

[Search criteria: skeletal muscle tissue histology morphology myofibrils stratum]

2.2.3.2 Rhythmic Contraction

The heart musculature exhibits some peculiarities, due to the central position of the heart in the circulatory system. For this reason it is fairly easy it to differentiate diagnostically. Like the skeletal musculature, it shows striations in a longitudinal section (Fig. 2.32). The striations in the heart also originate from the positioning of the Z-disks between the sarcomeres of different myofibrils, in the same plane.

A substantial difference to the skeletal musculature, however, is that heart muscle cells have only one centrally located nucleus. It is surrounded by a small myofibril-free area, in which organelles and granula are found. The substantial capillarization between the heart muscle cells, whereby continual pumping is made possible, is striking in its morphology. Furthermore, one finds few fibrocytes and only occasional collagen fibers between the heart muscle cells.

Nerve fibers are almost completely missing in heart muscle tissue. However, the heart has a special conducting system and pacemaker. These special heart muscle cells are richer in sarcoplasm and poorer in fibrils, in contrast to the myocytes of the skeletal musculature. They are also very rich in glycogen. These cells have the

Fig. 2.32: Histological representation of heart muscle. The branched cells, the centrally located nucleus and the intercalated disks as connecting elements are typical characteristics.

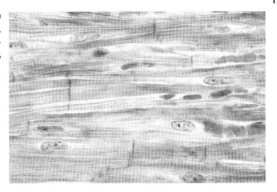

capacity to produce spontaneous and rhythmic action potentials. These action potentials then spread over the heart musculature, during systole, from the apex to the remaining ventricular musculature. The progressive propagation of the action potential is the result of special cell–cell connections, in the form of gap junctions, between the individual myocytes (Fig. 2.33). Together with the zonulae adhaerentes and desmosomes, which both provide mechanical connections, they are recognized under light microscopy as intercalated disks.

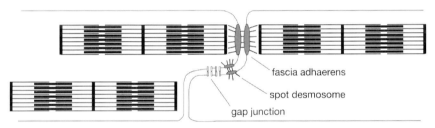

fascia adhaerens

spot desmosome

gap junction

Fig. 2.33: Schematic illustration of the contact zone between two cardiomyocytes. For mechanical and functional coupling, the fascia adhaerens, a spot desmosome and a gap junction are represented.

[Search criteria: heart muscle tissue histology morphology myocardium]

2.2.3.3 Unconscious Contraction

Smooth muscle is found in many internal organs and blood vessels (Fig. 2.34). It causes the gallbladder and the bladder to empty. In the intestine, smooth muscle is responsible for the oscillating motions that further the transport of food (peristalsis). In blood vessels it regulates the width of the lumen and thus the circulation in the organ or tissue.

The smooth muscle cannot be consciously controlled. As with heart muscle, and in contrast to skeletal muscle, it is controlled by the autonomic nervous system, through the sympathetic and parasympathetic networks. The nerve fibers form local swellings, or varicosities, at smooth muscle cells. Electrical coupling occurs across the gap junctions between smooth muscle cells, whereby peristaltic waves of contraction develop in the organ vessel wall.

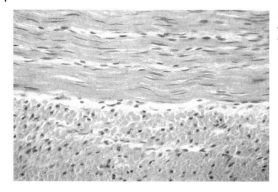

Fig. 2.34: Microscopic representation of smooth muscle – longitudinal and transverse section.

In addition to the contractile filaments actin and myosin, one also finds intermediate filaments in the cell (Fig. 2.35). Actin and myosin filaments are responsible for the actual contraction of the smooth muscle cell. However, they do not form the closely connected repetitive structural units, like the sarcomeres of the skeletal musculature. Rather, the actin filaments are disordered, connected to each other and to intermediate filaments by transverse cross-linking zones. Beyond that, a connection exists between the actin and intermediate filaments and the cell membrane. Thus, striations are missing in the smooth muscle cell.

Smooth muscle cells are up to 800 μm long and are mostly spindle shaped. Their relatively small nuclei lie centrally in the cell and are cigar shaped in their relaxed state. In the contracted state, they exhibit a typical corkscrew shape, as the nucleus is constricted. Smooth muscle cells do not possess their own extracellular covering. Rather, they are much more structurally, and thus functionally, integrated into a tissue or organ.

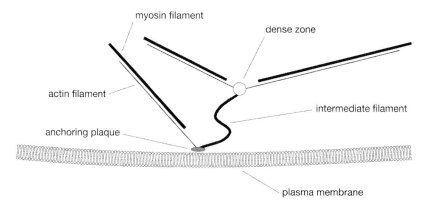

Fig. 2.35: Schematic illustration of contractile elements in a smooth muscle cell. Note the diffusely distributed actin and myosin filaments, which are anchored to the plasma membrane, over intermediate filaments.

[Search criteria: smooth muscle tissue histology morphology contraction]

2.2.4
Nervous System Tissue

Nervous system tissue develops out of the neuroectoderm and is thus a specialized epithelial tissue. It consists of nerve cells (neurons) and neuroglia (special neural connective tissue cells), and makes the exchange of information within an organism and between different tissues possible. The neurons form an informational and connective network, over a multitude of cell runners (dendrites and axons) (Fig. 2.36). A central and a peripheral nervous system are differentiated. Critical hormones are formed within particular areas of the central nervous system, as in the hypothalamus and the pituitary gland.

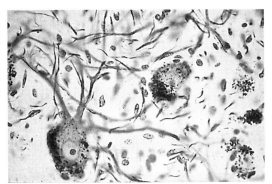

Fig. 2.36: Microscopic representation of the cerebellar cortex. Nervous tissue is differently developed in the individual areas of the central and peripheral nervous systems. The central connecting elements are the neurons. A multipolar nerve cell with numerous dendrites and one axon is representative.

[Search criteria: neural tissue histology morphology neurons glia]

2.2.4.1 **Information Mediation**

It is not within the scope of this text to describe all the characteristics of the central and peripheral nervous systems in detail. For this purpose it is advisable to refer to the extensive literature on the topic of microscopic anatomy. The peripheral nervous tissue primarily serves information transfer in the human body (Fig. 2.37). It generates or takes up excitation, which is passed on to other nerve cells or effector cells and tissues. In order to transfer information, nerve cells exhibit special cell runners. Thus, the cells are functionally polarized. On one side, the cell extensions are dendrites, which take up the action potentials. The action potential is passed on by a sequential membrane depolarization of the axon, which, in some nerve cells, is up to 1 m long. At the end of the axon, the neuron is connected by a synapse to other cells, by which it passes the excitation on. Extensions from other neurons can end at dendrites, forming axodendritic synapses. Thus, the communication networks are form.

Fig. 2.37: Schematic illustration of a multipolar nerve cell with several dendrites, perikaryon and an axon.

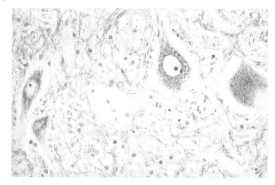

Fig. 2.38: Histological representation in the ventral root of the grey substance of the spine. Multiple cell bodies of a-motor neurons are shown. Between them are numerous nerve fiber connections.

The nerve cell body (soma) contains the nucleus and the synthesis apparatus, which produces the neurotransmitter for the transmission of the action potential (Fig. 2.38). Structurally, the cytoplasm of the dendrites is similar to that of the soma. The axon, on the other hand, does not possess portions of this synthesis machinery. This is already recognizable, under light microscopy, in the hillock area where the axon begins. Here, the Nissl granules are missing, when prepared with a special Nissl stain for highlighting this synthesis machinery. Under light microscopy, Nissl granules are visible portions of the ER, along with free ribosomes. The missing synthesis apparatus in the axon is replaced by special transport structures. Molecules and vesicles formed in the soma are transported by the motor protein kinesin along the microtubules. The vesicles filled with transmitters are then emptied at the synapse. The emptied vesicles again are returned to the soma by the motor protein, dynein.

Nerve cells can exhibit very different forms. The most frequent nerve cell is the multipolar nerve cell. It possesses several cell extensions, whereby only one of these is an axon. The other cell extensions are, accordingly, dendrites. Nervous system tissue, on the other hand, have only one dendrite and one axon.

Nervous tissue does not only consist of a neural network, but also of macroglia and microglia. The macroglia are divided into astrocytes, microglia, oligodendrocytes and microgliacytes.

Astrocytes are star-shaped cells in the central nervous system, with multiple cell extensions. They cover blood vessels in the central nervous system, are thus a component of the blood–brain barrier, thereby controlling the composition of the extracellular environment of the nervous tissue. Astrocytes can also produce very long cell extensions, forming connections between pyramid cells in the cerebral cortex and blood vessels.

Oligodendrocytes partly replace the ECM in the nervous system by covering the perikaryon and extensions of nerve cells, forming the myelin sheaths. The microglia are relatively small cells with oblong nuclei. They phagocytose particularly well, playing a substantial role in the repair process, and therefore plasticity, in the central nervous system.

Myelin sheath-building glial cells in the peripheral nervous system are the Schwann cells, whereas the oligodendrocytes perform this function in the central nervous sys-

Fig. 2.39: Schematic illustration of an axon (A) with a myelin sheath formed by a Schwann cell.

tem. Both cell types build myelin around the axon. Electrical insulation of the axon is the result. An axon with a myelin sheath is called a nerve fiber. During mediation, the Schwann cells wrap themselves repeatedly around the axon and cover it for a length of approximately 1–2 mm (Fig. 2.39). Myelin-free sections, referred to as nodes of Ranvier, exist between the Schwann cells (Fig. 2.40). The excitation is propagated by jumping from one node to the next. Myelination serves to significantly increase the impulse speed of transmission in the nerve. This process is called saltatory conduction. Schwann cells always surround only one axon; oligodendrocytes, on the other hand, can surround several axons. However, not all axons in either the peripheral or central nervous systems are myelinated. Furthermore, axons may be myelinated to different degrees.

Ependymal cells are epithelial glia cells. They line cavities of the central nervous system and have kinocilia on their surface, which cause movement of the liquor.

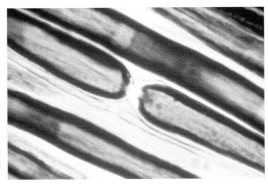

Fig. 2.40: Histological representation of a node of Ranvier, which lies between the myelin sheaths of two neighboring Schwann cells, after osmium contrast staining.

[Search criteria: peripheral neural tissue histology morphology axon]

2.2.4.2 Networks and Connections

White and grey matter can be differentiated in the central nervous system. The cell extensions of neurons and glia cells are in the white matter; the perikaryon of the neurons, as well as glial tissue, are in the grey matter. This structure is particularly recognizable in the cerebellum and cerebrum. The outer layer of both consists of grey matter and is referred to as the cortex, whereas one finds only the nerve fibers of the white matter inside the cortex or medulla. This distribution is also found in the spinal

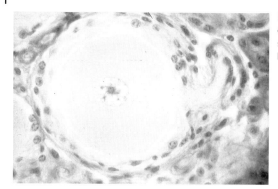

Fig. 2.41: Histological representation of a pseudounipolar nerve cell in the spinal ganglion, which is surrounded by numerous mantle cells.

cord. However, here, the grey matter with the perikaryon is found in the center and the white matter surrounds it. In the ventral horn of the grey matter one finds numerous multipolar nerve cells, which are described according to their function as motor neurons and are responsible for the innervation of the skeletal musculature of the trunk.

Individual axons leave that spinal cord from the ventral root and merge with peripheral nerves. In a histological cross-section of a peripheral nerve, many axons are recognizable, but never the cell bodies of neurons. The visible nuclei belong to the Schwann cells. A peripheral nerve does not only contain the motor fibers, going into the periphery, but also sensory fibers. They bring stimuli from the entire periphery to the spinal cord. The perikaryon of the sensory fibers lie in the spinal ganglia, where they are surrounded by special satellite cells (Fig. 2.41). The cell bodies that exist there belong to pseudounipolar nerve cells, since only one cell extension leaves the cell body and then divides in a T-shape. The fibers that run toward the spinal cord pass through the dorsal root, where they synapse with other neurons or continue on toward the brain stem. Peripheral nerves contain additional connective tissue. The endoneurium surrounds individual nerve fibers, while the perineurium surrounds multiple nerve fibers as a bundle. The epineurium covers the complete nerve as taut connective tissue.

As an important integration organ, the cerebellar cortex, should be briefly presented (Fig. 2.42). All information about coordination and fine-tuning of motor function, as

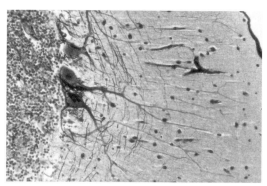

Fig. 2.42: Histological representation of the cerebellar cortex with the stratum moleculare (left), the stratum ganglionare, with the darkly colored Purkinje cells, and the stratum granulosum.

well as the regulation of muscle tone, reaches the Purkinje cells in the stratum gang-lionare. Directly neighboring the darkly colored Purkinje cells, in the stratum gran-ulosum, are the Golgi cells and granule cells. Stellate cells and basket cells are found in the stratum moleculare. The size of the excitation of the connected cerebellar nuclei is established in a complex automatic control loop in all of these cells.

[Search criteria: central nervous system neural tissue histology morphology]

2.3
Relevance of the ECM

2.3.1
Components of the ECM

The ECM is also referred to as intercellular substance. It consists, to a large extent, of fibrillar proteins, which form fibrous structures including various differently devel-oped collagen, reticular and elastic fibers. Histologically, the intercellular substance also contains another light-microscopic homogenous mass, the ground substance. This is formed mainly by the connective tissue cells. Depending on its biochemical composition, it is either solid or more gel-like and strongly hydrated. The predomi-nantly water-soluble components are extracted with the histological processing and therefore are usually not visible. The ground substance is of greatest importance in the selective exchange of material between the cells and the blood. The transport of nutrients and metabolic products also takes place it. A regulating effect on the trans-port of substances comes from the shift of the cell gel condition and the change in hydration.

2.3.1.1 **Functions of the ECM**
Hopefully, it has been illustrated that a relatively close spatial relationship exists be-tween the cells in the different tissues, and that different kinds and quantities of ECM and interstitial fluid are present in the respective intercellular spaces. The geometry of the intercellular space is can differ and fulfills specific tasks. Along the lateral borders of epithelia, for example, are very narrow, fluid-filled areas, whereas with the connec-tive and support tissues, large quantities of mechanically loadable intercellular sub-stance is present. With relatively thick areas of tissue, consisting of several cell layers, the intercellular space represents an important route of transport for nutrients and metabolites to and from cells. Epithelia and cartilage contain no blood vessels of their own, whereas all other tissues are highly vascularized.

For many years it was believed that the pericellular (or ECM) represented only a scaffold for cells and tissue. In recent years, however, it has become evident that a close structural–functional relationship exists between the ECM and the individual tissue cells, particularly between the cytoskeleton and the nucleus with its genetic information. The ECM has contact with cell surface receptors, which mediate signals from the outside into the cytoplasm. This, in turn, initiates a signaling cascade, which stimulates or inhibits internal regulatory systems and in this way influences gene expression in the nucleus. This can change either cell characteristics or the ECM. This is done via the increased synthesis or the dismantling of the ECM. This interactive process is called dynamic reciprocity. Adhesion, migration, cell division, differentiation, dedifferentiation and apoptosis can be controlled in tissue through this cellular mechanism.

The ECM consists of structural components, such as the different collagens, glycoproteins, hyaluronic acid, glycosaminoglycans, reticulin and elastin. Additionally, growth factors, cytokines, matrix-degrading enzymes and their inhibitors are stored in the matrix. A number of growth factors and cytokines interact with the ECM, stimulating a multitude of cell functions and thereby production or degradation of the ECM. Transforming growth factor-β (TGFβ), for example, can stimulate the formation of ECM components, simultaneously inhibiting enzymes such as metalloproteinases, which cause their break down. In addition, the ECM is not structurally uniform, but shows specialized characteristics in each tissue. Connective tissue cells are settled within a multitude of three-dimensional matrices, with special mechanical characteristics, whereas epithelial cells are anchored to planar basement membranes, with specific physiological functions.

Cells must bind to the individual components of the ECM. In addition, special areas are needed where cell receptors can anchor. For this purpose, ECM molecules have special motifs in their amino acid sequence which make receptor anchorage only possible at these positions. The best understood motif is the tripeptide RGD (arginine–glycine–aspartic acid). Among other responses, this sequence of amino acids stimulates fibronectin to adhere to cells. The same motif is found in laminin, entactin, thrombin, tenascin, fibrinogen, vitronectin, type I collagen I and VI, bone sialoprotein, and osteopontin.

Cell proliferation and subsequent differentiation occur three-dimensionally in the tissue. A prerequisite is the structured environment of the ECM, within which the cells can intensively interact with each other and the matrix. The ECM is comprised of different kinds of fibrillar macromolecules, to which the different collagens, elastin, fibrillin, fibronectin and proteoglycans belong. These protein molecules are woven into each other. Connections with hyaluronic acid and proteoglycans are also made.

Electrolytes and water are also found in the meshwork. The composition of the ECM varies according to the tissue type. Thus, the compression and tensile strength, as well as elastic deformation, can be adapted to the respective demands in the tissue. Components of the ECM are developed first in a very provisional form. These components are then dissolved by proteases in order to build in new components. Only through such on-going degradation does an ECM finally appear, as it is found in mature, adult tissues. With age, the composition of the ECM still changes continuously. Indicative of this is the formation of wrinkles. Out of a taut subcutical connective tissue in youth, wrinkles can develop, in principle, over the whole body, but particularly around the face, breasts and bottom.

The reciprocal effect between cells and ECM plays a decisive roll in tissue development and wound healing. Only through constant communication between the individual cells and tissues with the surrounding extracellular environment can morphogenic fields as well as the anlage for organs and tissues be developed and maintained. In wound healing, analogous to embryogenesis, primary plugging through blood coagulation, inflammation reactions, development of granular tissue and three-dimensional restoration must be coordinated. It is through these developmental processes that cell adhesion, detachment, migration, proliferation, differentiation and apoptosis, as well as matrix building and dismantling, are interactively directed.

[Search criteria: extracellular matrix fibers function]

2.3.1.2 Synthesis of the Collagens

The development of the ECM happens first within the cell and ends in the extracellular space. This can be shown particularly well with the example of collagen synthesis by fibroblasts (Fig. 2.43).

Pro-α-polypeptides, containing a signaling sequence and rich in the amino acids proline and lysine, are formed on the polyribosomes of the rER. These polypeptides are then taken up into the rER cisterns, where the signaling sequence is split off. Through the enzymes peptidyl-proline hydroxylase and peptidyl-lysine hydroxylase, a hydroxyl group is attached to the amino acids proline and lysine. Then, a further modification takes place as the hydroxyl groups are glycosylated (a specialized sugar residue is attached). A particularly high degree of glycosylation is found in the basement membrane of epithelia.

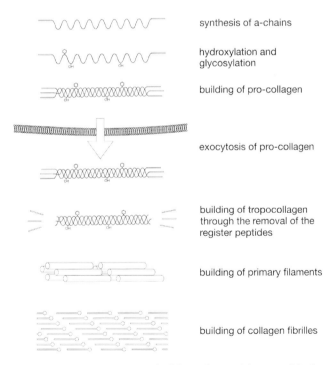

synthesis of a-chains

hydroxylation and glycosylation

building of pro-collagen

exocytosis of pro-collagen

building of tropocollagen through the removal of the register peptides

building of primary filaments

building of collagen fibrilles

Fig. 2.43: Schematic illustration of cellular synthesis and the extracellular formation of collagen fibrils.

The fibrillar collagens consist, in principle, of three subunits, the α-chains, which wind around one another to form a triple helix. Type II and III collagen are composed of same α-chains (homotrimers), whereas type I, V and XI collagen are composed of different chains. An α-chain is composed of approximately 1000 amino acids. In the next step, the α-polypeptides are twisted together into a triple helix.

Collagens are insoluble proteins. For this reason, additional amino acids that increase the solubility of the molecule, at the same time preventing aggregation, are already attached to the end of the α-chains during synthesis in the ER. Only in this way is it possible for the cell to remove the synthesized propeptide. In addition, register peptides ensure that the three chains attain the correct position and that the fibrillar triple helix remains soluble. The newly synthesized molecules are removed from the cell through the Golgi apparatus and exocytosis vesicles.

After the secretion of the molecule, the propeptide sequences attached to the C-terminal end are split off by a protease, while the propeptide on the N-terminal end is split off from the area near the collagen fibril. The resulting molecule at this point is tropocollagen. These molecules become insoluble through the splitting off of the register peptide. Microfibrils result through parallel aggregation and covalent cross-linking. Of importance are the aldehyde groups, which result from the enzymatic deaminization of lysine and hydroxylysine. This cross-linking is of crucial importance for the later tensile strength of the fibrils. The poor functional healing of an overstretched tendon can be attributed, mainly, to the unstable cross-linking of these fibrils during regeneration. Collagen fibers can form spontaneously through the aggregation of type I and type III microfibrils, whereas the aggregation mechanisms for other collagen types are still not known.

The collagen fibrils result, again, from the aggregation of several triple helix molecules. Collagen fibrils are heterotypically developed and can consist of more than one type of collagen. Thus, fibrils of collagen type I contain also collagen type V and collagen type II occurs together with type XI. This compositional relationship specifies whether thinner or thicker fibrils are produced. The ECM must finally offer a maximum in mechanical load with a minimum of building material. A prerequisite is that the ECM is perfectly three-dimensionally linked. Examples of this process are the fiber bundles in the lamellar bone and at the surface border between bones and hyaline cartilage.

[Search criteria: collagen synthesis fiber extracellular matrix formation]

2.3.1.3 Fibronectin

The connection between cells and the ECM is mediated by fibronectin. Fibronectin is a glycoprotein, comprised of dimers, each of approximately 250,000 molecular weight. Each of these subunits is folded three times and contains the amino acid repeats FN1, FN2 and FN3. Fibronectin can bind to collagen at the FN1/2 region, while the FN3 region takes up contact with the respective cells. The contact points consist of the amino acid sequence RGD and bind integrins. The α5β1-integrin has a particularly high affinity for fibronectin. Completely different kinds of fibronectins are formed by alternative splicing. The liver, for example, builds fibronectins which then circulate as a dissolved component of the serum. However, individual stationary fibronectins are also formed in the different tissues.

[Search criteria: fibronectin extracellular matrix function]

2.3.1.4 Laminin

Laminins are large molecules with a molecular weight between 140,000 and 400,000, which occur mainly in the basement membrane of the epithelium. They consist each of an α-, β- and γ-chain. From the aggregation of the chains, X- and Y-shaped molecules result. The laminins, on the one hand, can bind to other components of the ECM and, on the other hand, have binding sites for cell receptors. Several repeated amino acid sequences are present on the short arm of the laminin molecules that are also found in epidermal growth factor (EGF). Between, lie globular domains that, in the presence of calcium, contribute to the cross-linking of the basement membrane. The high affinity of nidogen for the laminins is particularly noticeable. Its binding site sits on the γ_1-chain, at the cross-center of the molecule. Another globular domain of the laminin molecule binds to type IV collagen and ensures further cross-linking of the basement membrane. In most cases, laminin binds only indirectly to type IV collagen. Other bridge molecules in addition to nidogen have been shown to be heparin, perlecan and fibulin-1.

[Search criteria: laminin extracellular matrix function]

2.3.1.5 Reticular and Elastic Fibers

Tissues and organs with a high content of reticulin and elastin have the characteristic that they can return intact to their initial position after a temporary mechanical load and thus are in the truest sense elastically deformable. Reticulin is a component of many lymphatic tissues and parenchymal organs, like the liver. Elastin is found predominantly in arteries near the heart, the skin and in the lung. Elastic fibers show about a five-fold larger elasticity than rubber. Microfibrils consisting of fibrillines are found on the surface of elastin fibers. Interestingly, these molecules contain repeats of the amino acid sequence in EGF and TGFβ.

[Search criteria: reticular elastic fibers extracellular matrix function]

2.3.1.6 Collagens of the Basement Membrane

The basement membrane, as a specialized form of the ECM, occurs at the surface between the epithelium and connective tissue. At least six different genes ensure that type IV collagen develops a planar branching network in the basement membrane. Type IV collagen is connected by a set of non-collagen proteins. These include the different isoforms of laminin, as well as nidogen and perlecan. Type XVII and VII collagens have been shown to be present in places of particularly large mechanical stress. Type XVII collagen is a transmembrane molecule that can fix cells to the fibrils of the basement membrane. Such binding points are found in hemidesmosomes. Collagen types XV and XVIII belong to the multiplexins (multiple triple-helix domains and interruption). These are found between the endothelium and the tunica intima in blood vessels, in particular in the basement membrane. If collagen XVIII is split by proteases, heparin-binding fragments are set free which prevent the production of new blood vessels. One of these fragments is endostatin and lies on the C-terminus of type XVIII collagen. If this peptide is synthesized, it can block the proliferation as well as migration of endothelial cells, bringing tumor growth to a stop.

[Search criteria: basement membrane collagen extracellular matrix function]

2.3.1.7 FACIT Collagens

In addition to the classical collagens, with a pure triple-helix structure, there are molecules in the ECM which contain other typical protein domains. Type IX, XII, XIV and XIX collagens belonging to this group. Type IX collagen, for example, is found on the surface of type II/XI collagen fibrils. The long part of the molecule lies parallel to the fibrils, while the short part ascends into the perifibrillar area. Due to its position, it is assumed that the type IX collagen molecule can make connections both to neighboring fibrils and to other molecules of the ECM. It is also well known that type IV collagen can bind both to heparin sulfate and decorin, which is again associated with collagen fibrils.

[Search criteria: FACIT collagen extracellular matrix]

2.3.1.8 Proteoglycans

Proteoglycans have varied tasks. Aggrecan and versican, as high-molecular-weight representatives, bridge wide areas in the ECM in cartilage, together with hyaluronic acid. Syndecan is localized in the plasma membrane and serves as a cell receptor. Perlecan is not only found in the basement membrane of the epithelium, but also in the pericellular matrix of other tissue cells. In the liver, perlecan is built by endothelial cells of the sinusoids. Syndecan is a transmembrane protein that can bind growth factors, protease inhibitors, enzymes and components of the ECM on the cell surface.

Small proteoglycans such as decorin, biglycan, lumican and fibromodulin interact with components of the ECM. Decorin, for example, binds to collagen fibrils, thereby playing a substantial role in the assembly of the collagen fibers.

Hyaluronic acid is found in nearly all ECM. It serves as a ligand for the cartilage link proteins, aggrecan and versican. However, cell receptors such as CD44 can also bind hyaluronic acid, and therefore affect cell proliferation and migration. Cell-free areas are created by hyaluronic acid deposits during tissue development. These areas are then opened up by hyaluronidase, allowing cells to migrate and condense in order to develop tissue structures.

[Search criteria: proteoglycan extracellular matrix]

2.3.2
Interactions between the Cell and the ECM

2.3.2.1 Adhesion and the ECM
The significance of the interaction between cells and the ECM can be impressively illustrated in the gastrulation of the embryo, migration of the neural crest cells, angiogenesis and building of the epithelium. The extracellularly occurring fibronectin interacts closely with early embryonic cells. Such cells, if injected with antibodies against fibronectin or peptides with RGD amino acid motifs during development, develop false formations. Cell movement may be partly reduced, bilateral symmetry may be missing or malformations of the circulatory system may develop. If integrin binding sites are blocked with the appropriate peptides, complete misformation in the gastrula and neural stages of development can be observed.

During angiogenesis or the development of new blood vessels from existing structures, individual proteins of the ECM are of the greatest importance for the migration of endothelial cells. The cells can independently form tubular structures, providing they are cultivated in a matrix, stemming from an Engelbreth-Holm-Swarm (EHS) tumor. This ECM contains type IV collagen, proteoglycans and entactin. The development of tubular structures is lacking, however, if the cells are kept in the presence of type I collagen. Obviously, laminin promotes the development of the endothelial tubuli. If endothelial cells are cultivated in the presence of both the EHS matrix and lamini antibodies, the tubulus formation is missing. The growth of endothelial strands cannot only be achieved with the intact laminin molecule, but also with a peptide containing the amino acid sequence SIKVAV, which is also found in the α-chain of laminin. The amino acid sequence CDPGYIGSR-NH$_2$ is found in the β-chain of laminin. If this peptide is used under culture conditions or in animal experiments, angiogenesis is blocked. Although the molecular sequences of these reactions are not known in detail, it is nevertheless surprising that both tissue development promoting and inhibiting motifs are present in laminin.

[Search criteria: cell adhesion extracellular matrix interaction]

2.3.2.2 Proliferation and the ECM
Based on laminin, it can be shown how a protein of the ECM can affect cell proliferation. Numerous EGF repeats are found in its α-chain, i.e. repeated amino acid sequences found EGF. These have an intensifying effect on the proliferation of numerous cell lines. Likewise, the increased proliferation of macrophages can be shown in the presence of laminin.

Heparin also shows inhibition of cell proliferation. Endothelial cells of the aorta, cultivated in a heparin-containing media, show no cell proliferation. After treatment of the medium with heparinase, the inhibition is lifted and proliferation begins. This effect cannot be obtained with chondroitinase or protease, which indicates the specificity of the test. Another example is with human mammary gland cells. These cells proliferate permanently if they are held in polystyrene-surfaced culture dishes. If, however, the dish is coated with proteins of the ECM, cell proliferation is inhibited. Finally, it can be shown in the cultivation of hepatocytes that the ECM inhibits the expression of immediate-early growth response genes and at the same time induces C/EBPα, which leads to the turning on of metabolic function genes.

For biological reactions between cells and the ECM, the cooperation of growth factors is frequently needed. Basic fibroblast growth factor (bFGF), interleukins (IL-1, IL-2, IL-6), hepatocyte growth factor, platelet-derived growth factor (PDGF)-AA and TGFβ are found in large quantities in the ECM, and frequently delivered elsewhere on demand. A close interaction between cells, the ECM and TGFβ can be shown in the development of the mammary gland. On the one hand, epithelial cells of the mammary gland need to multiply and branch out into ducts. This happens in close cooperation with the surrounding ECM. On the other hand, after the growth phase, the expanded mammary glands must differentiate, may no longer increase in size and must be stabilized with the surrounding connective tissue. TGFβ is now produced by the epithelium, which inhibits cell proliferation and the enzymatic dismantling of the ECM by the metalloproteinase stromelysin-1. It is noteworthy that TGFβ is not found within areas of completely new gland growth. Thus, enzymes for dismantling the ECM are not inhibited here. In this way, it is possible for the duct system of the mammary gland to spread out while new ECM is being formed.

[Search criteria: proliferation control extracellular matrix interaction]

2.3.2.3 Differentiation and the ECM

The ECM has a decisive influence on the differentiation of tissue. The stratified squamous epithelium of the skin is formed by keratinocytes. Within 30 days the epithelial layer is renewed by the stratum basale. In order to do this, the stem cells in the stratum basale must always be multiplying. In this phase, they have direct contact to the basement membrane and do not show the typical expression of proteins found in terminally differentiated keratinocytes. The regeneration of the epidermis is based on numerous asymmetrical cell divisions in the stratum basale, since many cells separate from the basement membrane and migrate to the suprabasale cell layer. During this first developmental step, the first differentiation markers, such as involucrin, become visible. In this example, it is clear to see that a new development program is switched on with the replacement of these cells by the basement membrane. This is confirmed by culture experiments. Cells of the stratum basale that are isolated and kept in suspension culture without the ECM develop in a much shortened differentiation program. In order to experimentally examine the development of stratified squamous epithelia *in vitro* under the most physiological conditions possible, keratinocytes are frequently cultivated on a layer of 3T3 cells, so that the synthesis of a basement membrane and the natural differentiation program, typical for the tissue, are supported. Many patients with severe burns have been successfully treated using this method over many years.

With hepatocytes, differentiation typical to this tissue is observed when they are cultivated on an EHS matrix. Three transcription factors can be shown, eE-TF, eG-TF/HNF-3 and eH-TF, which are only activated if cells are in an EHS matrix with sufficient laminin. There are similar findings for the epithelia in milk glands, which do not express tissue-typical proteins in normal culture dishes. If, however, the cells are grown on an EHS matrix, they develop alveolar structures and begin the expression of typical milk proteins, such as β-casein. Two different processes can be observed with this development. The first result is a shape change and, thus, changes in the cytoskeleton. Second, a tyrosine kinase signal is activated by the β1-integrin receptor, which ultimately leads to the formation of β-casein. This signaling cascade, transmitted from the ECM and the β1-integrin receptor into the cell, is also observed in the synthesis of albumin in hepatocytes.

[Search criteria: cellular deviation control extracellular matrix interaction]

2.3.2.4 Apoptosis and the ECM

In embryogenesis, programmed cell death (apoptosis) belongs to the normal development phenomena. Apoptosis is particularly impressive to observe during the maturation of the blastocyst, development of the extremities, the palate and the nervous system, and during thymocyte differentiation, development of the mammary gland and in the development of vascular structures. It has been shown, for example, that during the development of the mammary gland at the end of the pregnancy, the ECM suppresses apoptosis of the epithelia. After breast-feeding has come to an end, milk-producing alveoli, along with the corresponding ECM, are dismantled. The loss of the cell–ECM interaction is accompanied by a rise in caspase-1 activity, which in return supports apoptosis. If the binding of β1-integrin to the ECM is blocked with an antibody, this also leads to apoptosis in the mammary gland tissue. Similar cell–ECM interactions can be found in angiogenesis. If the integrin binding to the endothelial cells is interrupted here, the development of new blood vessels is omitted.

[Search criteria: apoptosis control extracellular matrix interaction]

2.3.3
Signal Transduction

2.3.3.1 Modulation of the Cell–Matrix Interaction

While fibronectin and laminin exclusively support the adherence of cells to the ECM, other neighboring molecules, such as thrombospondin and tenascin, can modulate this interaction positively or negatively. Thrombospondins contain multiple EGF repeats and calcium-binding sites. Thrombospondin1 is formed by fibroblasts, endothelial cells and smooth muscle cells, and binds fibrillar collagen, fibronectin, laminin and heparan sulfate proteoglycans. It shows growth-promoting effects with fibroblasts and seems to destabilize the cell–matrix interaction. Thus, it supports cell proliferation and angiogenesis. Analogous functions are found in cartilage. The cartilage oligomeric matrix protein (COMP) has molecular similarity to the thrombospondins. It is made by chondrocytes and secreted in the pericellular matrix. Either too little or missing synthesis of COMP leads to the softening of the otherwise mechanically stable cartilage matrix.

The tenascins consists of three (tenascin-X) or six (tenascin-C/R) subunits. Repeats for type III fibronectin and EGF-like domains, and binding sites for β- and γ-fibrinogen chains are found in the amino acid sequence. Tenascin expression is tissue specific. Tenascin-R is formed during development of the nervous system, whereas tenascin-X is found specifically in the smooth, heart and skeletal musculature. Tenascin-C, on the other hand, is found in healing wounds, in many tumors and in the brain. This molecule can support the attachment of cells over receptors and proteoglycans. Its effect can be inhibited by interaction with fibronectin. Cell surface molecules, such as contactin, react with tenascin-C/R and can stimulate or inhibit the growth of axons during development of neurons.

[Search criteria: cell extracellular matrix interaction signaling]

2.3.3.2 The ECM and Cell Binding

The long-term functional attachment of cells to the ECM takes place over cell receptors, which bind to special amino acid motifs in the ECM. Such contacts are mediated by molecules of the integrin family, which are localized in the plasma membrane of the respective tissue cell type. These are transmembrane proteins consisting of two units (dimers). Each of these is composed of one α and one β subunit. Due to the multiple subunits, many different combinations can define the specificity of the anchoring receptor. These, in turn, correspond to particular sequences in the ECM. Experimental data show that some integrin receptors very specifically bind to an individual motif in the ECM, while others can bind to multiple motifs. Thus, a substantial plasticity is developed, and this explains why that cells and tissues are able to develop on very specific, as well as completely non-specific, artificial matrices.

Integrins can manufacture contact not only between a cell and the ECM, but also between neighboring cells. The $\beta1$- and $\beta3$-integrins mainly mediate the connection between a cell and the ECM, while $\beta2$-integrins are involved in cell–cell contacts. The $\beta1$-integrins are commonly found with connective tissue cells, and seek contact to fibronectin, laminin and collagen. The $\beta3$-integrins within the vascular system, on the other hand, show binding to fibrinogen, von Willenbrand's factor, thrombospondin and vitronectin.

Integrins are transmembrane components of the plasma membrane, but are not firmly connected. The mechanically stable connection between a cell and the ECM is realized over two positions. The integrin appears like a pin, which is pushed through the plasma membrane. The molecule does not sit symmetrically, but consists of a long external portion and a short internal portion. The large extracellular domain protrudes from the plasma membrane, binds divalent cations and then to an amino acid motif of a protein in the ECM. The smaller intracellular domain, on the other hand, interacts with the cytoskeleton of the respective cell. Cell biological information is thus conveyed from amino acid-binding motifs over the individual integrin molecule to the cytoskeleton. Cell shape, growth and differentiation can be influenced through modulation of this connection.

Integrins are heterodimerically built, each molecule consisting of one α and one β subunit, and not covalently bound to another. The subunits consist in each case of a large extracellular and a small transmembrane domain. With integrins on the cell surface, a cell can recognize another cell or the ECM. Eighteen homologous α and eight homologous β subunits have been described, which can in turn form more than 20 different heterodimers. Therefore, different cells will also have different sets of integrins. Leukocytes, for example, express β2-integrins such as αLβ2, αMβ2 and αxβ2, while β1-integrins are seen in many tissue cells. The binding ligands are, likewise, variously developed (Tab. 2.2). Many integrins bind the RGD motif found in fibronectin, vitronectin, fibrinogen and the von Willebrand factor. In contrast, β1-integrins bind ICAM1, ICAM2 and ICAM3, which do not contain the RGD sequence.

When integrins make contact to other cells or the ECM, this information is passed on to the inside of the cell, which activates a signaling cascade (Fig. 2.44). If a β1-integrin passes this information from the exterior of the cell to the interior, focal adhesion proteins, paxillin, talin and FAK take part in the signaling cascade. This causes more FAK to be phosphorylated, causing proteins such as Src to bind. This again causes both the autophosphorylation of FAK and the phosphorylation of other proteins, such as paxillin and tensin, whereby the focal adhesion point is connected to the cytoskeleton. At the same time, c-Jun N-terminal kinase (JNK) and extracellular regulated kinase (ERK) are activated via p130CAS and mSOS, respectively The α-integrin subunit stimulates Fyn and the membrane protein caveolin. These different signals cause integrins to be able to stimulate such events as growth or the development of a phenotype. At the same time, apoptosis is inhibited as the anti-apoptotic protein bcl-2 is highly regulated.

The enormous significance integrins have in the mediation of signals between the cell exterior and interior becomes obvious, and thus it becomes understandable that integrins are also crucially importance for tissue engineering. As cells are settled onto

Tab. 2.2 Examples for the selective binding of integrins to proteins of the ECM.

Connection to the ECM	Integrin
Laminin, collagen	α1β1
Laminin, collagen, fibronectin	α2β1
Laminin, collagen, fibronectin	α3β1
Fibronectin, vascular cell adhesion molecule-1	α4β1
Fibronectin	α5β1
Laminin	α6β1
Laminin	α7β1
Laminin, collagen, fibronectin, vitronectin	αvβ1
Factor X, fibronectin, complement protein C3bi	αxβ2
Complement protein C3bi	αMβ2
Intercellular adhesion molecule-1/2	αLβ2
Von Willebrand factor, laminin, fibronectin, vitronectin, thrombospondin	αvβ3
Laminin, fibronectin, vitronectin, thrombospondin, fibrinogen	αIibβ3

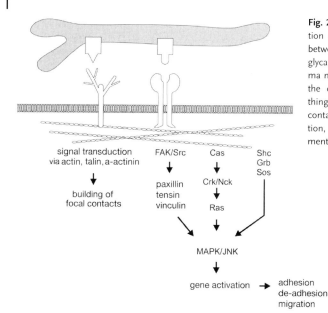

Fig. 2.44: Schematic illustration of the functional coupling between ECM proteins, proteoglycans and integrins in the plasma membrane. The coupling of the cell causes, among other things, the development of focal contacts, special gene activation, further adhesion, detachment and migration.

a scaffold, integrins first mediate, by recognition of their environment, whether a durable adhesion is present and whether a functional tissue can develop through these interactions.

In addition to integrins, proteoglycans sit transmembranously in the cell and can also adhere to the ECM, e.g. syndecan, CD44, RHAMM (receptor for hyaluronate-mediated motility) and thrombomodulin. Syndecan couples the cells to the ECM over chondroitin sulfate and heparan sulfate glycosaminoglycans. In contrast to integrin binding, this reaction is not calcium dependent. Syndecan is intracellularly bound to the cytoskeleton, allowing it to transmit information from the ECM into the cell interior. CD44 also carries chondroitin sulfate and heparan sulfate glycosaminoglycans on its extracellular domains. The binding site has six cysteine residues, which form three disulphide bridges, making this site very similar to the hyaluronic acid-binding site of aggrecans. RHAMM has been identified as a docking site for hyaluronic acid. Thrombomodulin is also a transmembrane protein, contains six EGF-like amino acid sequence repeats extracellularly and is functionally coupled via glycosaminoglycans.

In addition to the integrins and proteoglycans there exist proteins that can bind to the ECM via completely individual amino acid motifs. A laminin-binding protein recognizes a YIGSR sequence, which is not recognized by integrins. CD36 binds collagen, thrombospondin, as well as endothelial and some epithelial cells.

[Search criteria: cell adhesion extracellular matrix interaction receptors]

2.3.3.3 **Signals to the Inner Cell**

Molecules of the ECM react with cell receptors. This initiates a reaction cascade within the cell, via a second messenger, which can affect a variety of genes. Cell attachment, proliferation, migration, differentiation and death are affected through this mechanism (Fig. 2.45).

Integrins and proteoglycans are mainly involved in attachment, detachment and cell migration. If fibronectin, with its respective binding sites, connects at the same time to a suitable integrin and a proteoglycan, over the heparin-binding site, cell migration can be initiated. It has been shown that the receptors for proteoglycans are attached to intracellular microfilaments, which are connected again to integrin receptors. This coupling takes place within focal adhesions and in the direct vicinity of structural proteins, such as talin and α-actin filaments. The end of the integrin molecule in the cytoplasm is functionally coupled to a tyrosine kinase (focal adhesion tyrosine kinase). With the appropriate stimulus, phosphorylation of the enzyme takes place. In this reaction, c-Src (non-receptor tyrosine kinase), paxillin, tensin, vinculin and a further protein (p130) are phosphorylated. It is known that paxillin and tensin can transfer signals from the plasma membrane to the cytoskeleton via a phosphorylation. p130 interacts with further proteins, such as Crk and Nck, which can initiate cell migration via a mechanism in which Ras and MAP/JNK kinase are involved. Another control system is mediated by c-Src, FAK and the Grb2/Sos complex.

The connection of integrins to the ECM can be accelerated by molecules in the cytoplasm, such as cell adhesion modulator (CAR). The connection of the plasma membrane to the cytoskeleton can be intensified or lessened, e.g. the affinity between

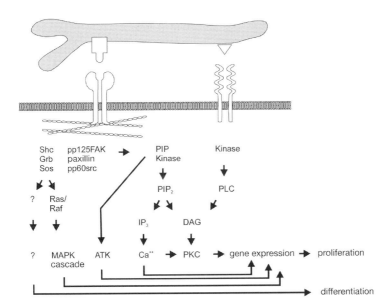

Fig. 2.45: Schematic illustration of the functional coupling between ECM proteins, receptors in the cell membrane and differentiation. The signaling cascade is initiated through complex reciprocal effects.

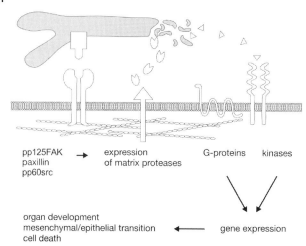

Fig. 2.46: Schematic illustration of the functional coupling between the ECM and tissue emergence. Components of the ECM are broken down through coupling with the cell, which sets free peptide fragments with morphogenic characteristics able to induce the location of tissue development.

pp125FAK $\rightarrow$ expression $\qquad$ G-proteins $\quad$ kinases
paxillin $\qquad$ of matrix proteases
pp60src

organ development
mesenchymal/epithelial transition $\quad\longleftarrow\quad$ gene expression
cell death

α2β1-integrin and collagen type I can be increased. Since the integrins and proteoglycans themselves do not have kinase or phosphatase activity, it is assumed that the signaling effect is transferred over regulatory proteins.

Apart from attachment and migration of cells, ECM–cell interactions affect differentiation. This cascade is activated if integrin receptors bind to the corresponding ECM, while growth factor receptors occupy their receptors on the cell surface. Thus an additional kinase [phosphatidylinositol phosphate (PIP) kinase] is activated, which increases the quantity of PIP2 and subsequently, phospholipase Cγ (PLCγ). Through further steps (PIP$_2$, diacylgycerol, IP$_3$) Ca^{2+} is set free from the ER, which leads to an activation of receptors on the cell surface and to gene expression. Based on this mechanism, it can be explained how cells bound to the ECM are much better able to react to hormone signals.

In addition, apoptosis and transitional areas of epithelial–mesenchymal differentiation can be directed via the ECM–cell interaction (Fig. 2.46). The signaling cascade for apoptosis is mainly initiated by type III collagen. Tyrosine kinase is also involved. With the epithelial–mesenchymal differentiation transitional areas, on the other hand, an increased degradation of the ECM takes place. Peptide fragments are set free, which can affect receptors on the cell surface, directing differentiation. Fragments of fibronectin can, in this case, bind α5β1-integrins, thereby activating them.

[Search criteria: signal transduction pathway extracellular matrix receptors]

2.3.3.4 The ECM and Long-term Contact

The ECM is a three-dimensional net composed of proteins and glycosaminoglycans. On the one hand, it facilitates the mechanical attachment of cells. On the other hand, important cell biological information is transmitted over the temporary or permanent attachments into the inner cell. In this way it is communicated to a cell whether it should migrate or stay in place in order to develop specific functions. The interaction is mediated mainly by glycoproteins or proteoglycans of the cell surface and the amino acid sequences of the ECM proteins.

Belonging to the family of the cadherins are surface receptors that make possible contact with homogenous cells (homophilic contacts). This happens with homogenous epithelial cells and requires the presence of Ca^{2+} ions. If the extracellular Ca^{2+} is extracted from culture medium by a chelating agent, such as EDTA, these contacts are dissolved, so that cells become detached and can be isolated as single cells. The connection between cadherins and the ECM is apparently only of lesser importance.

Selectins are also membrane proteins that mediate the contact between different cell types (heterophilic) in the presence of Ca^{2+} ions. They possess lectin-like characteristics and recognize the short-chain sugar molecules of their connection partners (sialyl Lewis-X/A).

Cell adhesion molecules (CAM) can have both homophilic as well as heterophilic characteristics, and thus take up contact to homogenous and heterogeneous cell types. Their connection to adjoining cells is Ca^{2+} independent.

While cadherins, selectins and CAMs almost exclusively mediate cell–cell contact, integrins can additionally mediate connections between the cell and the ECM. The β2-integrins are mainly involved in cell–cell recognition, while the β1- and β3-integrins manufacture contact between the cell and the ECM. The β1- and β3-integrins can bind to a whole set of ECM proteins, such as collagen, fibronectin, vitronectin and laminin.

Collagen is the most common protein of the ECM. Many different collagens exist and a completely different set is expressed in each tissue. A wide variety of receptor molecules can bind the collagens. Integrins α1β1 α2β1 and α3β1 particularly bind collagen. Fibronectin, which exists in many different variants, is frequently interlaced with collagen. Nearly all cells interact with fibronectin through the α5β1-integrin receptor, but there are also very specific receptors such as the αvβ3 receptor. Vitronectin is a multi-functional adhesion protein that can bind many cell types and through the vitronectin receptor αvβ3, αvβ1 and αIIbβ3 (blood platelet receptor). von Willebrand factor is formed by megakaryocytes of the marrow and stored in the α-granula of the circulating blood platelets. This factor is also produced by endothelial cells. Only about each tenth molecule is built into the subendothelial layer of the vasculature in insoluble form. Following damage to the vasculature, platelets can aggregate at this factor. Laminin is a complex adhesion molecule found in the basement membrane. It is a high-molecular-weight protein that is able to bind a variety of different integrins, whereby epithelia, mesothelia and endothelia are firmly anchored onto the basement membrane.

Since numerous molecules are involved in the structure of the ECM, it must be prevented from falling apart and its mechanical stability ensured. For this reason, cross-linking of the extracellular proteins takes place through transglutaminase activity. At the same time, there are neighboring areas of the ECM that are built up and dismantled by proteases. Newly synthesized ECM molecules must be built into these spaces so that the integrin receptors of cells settled over, or within, the ECM find suitable binding sites on collagens, glycosaminoglycans, fibronectins and laminins.

In most cases, the binding sites of integrin receptors within the ECM molecules have been shown to consist of oligopeptide sequences, which are comprised of up

Tab. 2.3 Binding domains of cell receptors on the ECM. Matrix proteins contain specific information sequences that are recognized by individual tissue cells and used for adherence.

Sequence	Protein	Function
RGDT DGEA	type I collagen	adhesion of many cells
LRGDN YIGSR PDSGR	Laminin	adhesion of epithelial cells
RGDS LDV REDV	fibronectin	adhesion of many cells
RGDV	vitronectin	adhesion via integrine αvβ3
RGD	thrombospondin	adhesion of many cells

to 10 linear or repetitive amino acids (Tab. 2.3). One of the most well-known sequences is the RGD motif, which was discovered in the fibronectin molecule and can bind numerous integrins. During cell biological investigations it has been shown that cell binding is very specific and only occurs at the RGD motif if the correct sequence is present. If the order of the three amino acids is changed, then no connection will take place.

Aside from the highly specific connection to the peptide sequences of the ECM proteins, cell surface molecules can also bind via less specific mechanisms. This takes place over heparin-binding domains, whereby proteoglycans containing heparin or chondroitin sulfate are recognized by the cell surface. Typical examples are the cell–cell adhesion molecules, in particular the neural cell adhesion molecule (NCAM), which possesses the binding domain KHKGRDVILKKDVR.

Cell biological reactions are initiated during the contact of a cell with the ECM. This must be understood as a bidirectional process. The cells accept cell biological information which comes from the ECM, but the matrix is arranged by the cells, around and corresponding to, the structure. Metalloproteinases secreted by the cells, such as collagenase, gelatinase, serine proteases, cathepsin and plasmin, are particularly significant. Space is created by these enzymes where synthesized protein, such as fibronectin, can be built into old fibrillar collagen structures. Thus, again, cells can migrate to this area and arrange the structure of new components. These processes have special cell biological significance in the functional adjustment of growth factors, which are only meant to be temporarily active, within a clearly defined framework. Many of the growth factors, such as bFGF or vascular endothelial growth factor (VEGF) bind with high affinity to heparin, and can thereby be bound within the ECM. As long as the growth factors are bound, they cannot trigger biological activity. They will only become biologically active if the matrix is dissolved by proteases or, as with VEGF, parts of the growth factors become enzymatically set free.

At the first contact of a cell with the ECM, focal contacts develop in which an increased concentration of integrins can be shown. This integrin clustering causes an increased tyrosine phosphorylation by proteins, particularly the pp125[fak] (pp125 focal adhesion kinase). Since the cytoplasmic side of integrins does not have catalytic activity, the mechanism by which the signal transmission takes place is still unknown. The signal transmitted into the cell can affect cell division or differentiation. In cultivated hepatocytes it has been shown, for example, that coating of growth surfaces with small concentrations of fibronectin or laminin stimulates the synthesis of albumin as a differentiation marker, whereas high concentrations of ECM proteins inhibit the synthesis of albumin and stimulate cell division. With cultivated neurons, it has been shown that by coating the culture dish with laminin, neurites proliferate more than when using fibronectin. It is clear that the development of new biomaterials for use as artificial ECM and the engineering of functional tissues will only be successful if cell biological interactions can be carried out as under natural conditions.

[Search criteria: cell extracellular matrix interaction]

2.3.4
Matricellular Proteins

The extracellular environment is of great importance, not only for cells, but, in particular, for the development of tissues. Included, apart from growth factors, is the relationship of a cell to its neighboring cells, its continuous contact with the ECM and the matricellular proteins cells that belong to it. All these components regulate the interaction of the cell's surface activity, the intracellular signaling cascades and, thus, gene expression. This, in turn, leads to cell migration, differentiation, and, therefore, to the formation of socialization and complex tissue structures. Matricellular proteins, which are found in secreted form in the ECM, but are not structural components (Tab. 2.4), are of great importance. It is possible that matricellular proteins are modulators that

Tab. 2.4: Examples of the cooperation of matrix cellular proteins with components of the ECM and cell receptors.

Protein	ECM interaction	Receptor	Modulation
Thrombospondin	type I, V collagen Laminin Fibronectin Fibrinogen	integrin CD36	HGF (−) TGF-b (+)
Tenascin C	fibronectin	integrin annexin	EGF (+) bFGF (+) PDGF (+)
Osteopontin	type I–V collagen Fibronectin	integrin CD44	?
SPARC	type I, III, IV, V collagen	?	EGF (−) VEGF (−) PDGF (−) TGF-b (+)

mediate the signals between cytokines, proteases, the ECM and cell receptors. Included in this group of proteins are thrombospondin-1, thrombospondin-2, tenascin C, osteopontin and SPARC (secreted protein, acidic and rich in cysteine).

2.3.4.1 Thrombospondin

Thrombospondins are high-molecular-weight macromolecules of about 450,000. The molecule is inserted into the plasma membrane in such a way that it can bind to components of the ECM, such as type I and V collagen, laminin, fibronectin, fibrinogen, and SPARC with at least five extracellular domains. On the other hand, contacts to integrin receptors of the cell can be developed. This interaction can be disturbed if, for example, endothelia and smooth muscle cells are cultivated with thrombospondin antibodies. In this case, no new tubular vesicular structures can form. Thrombospondin can also affect the attachment and, therefore, the form of endothelial cells. Cell migration can be initiated as necessary through stronger or weaker binding to the ECM. This mechanism has significance in the wound healing of the skin, but also in the migration of metastasizing tumor cells.

2.3.4.2 Tenascin C

Tenascin C is found in higher amounts during development and in lesser amounts in functional tissues. Frequently, the molecule is coupled to fibronectin in the ECM, whereas cellular contacts are put together with at least five integrins and annexin II. The strength of attachment is apparently dependent on the profile of the receptors just developed in the cell. EGF and bFGF are involved in this process. This reciprocal effect can be shown very impressively in smooth muscle cells, cultivated in a collagen matrix, and which secrete metalloproteinases. Binding sites for integrins on the cell surface are opened up through the dismantling of collagens. Tenascin C, which is secreted and stored in the ECM, is formed in response to the binding of integrins to collagen. Tenascin C then serves as a further ligand for the cell integrins. This causes focal contacts to be reorganized. Simultaneously, an increased development of EGF receptors can be observed on the opposite side of the cell, which again affects the cellular proliferation rate.

2.3.4.3 Osteopontin

Osteopontin is found not only in bone structures, but also in a variety of tissues. The molecule is bound to type I, II, III, IV and V collagens in the ECM, and numerous integrins along with CD44 from the cell side can dock onto it. If the synthesis of osteopontin is inhibited in smooth muscle cells by osteopontin at a molecular biological level, this leads to a worsening of cell attachment and to an increased propagation in artificial matrices. Binding sites for integrins can be opened on the molecular surface by the protease thrombin. As a result, more integrin receptors bind, which causes more integrin to be formed. Osteopontin is apparently of particular importance during the maintenance of differentiation. This phenomenon can be shown with endothelial cells, which depend on the presence of growth factors in the culture medium. In the absence of growth factors, apoptosis begins in the cells. However, cell death can be prevented by cultivation of the cells on an osteopontin substrate.

2.3.4.4 SPARC

SPARC (also BM-40, osteonectin) was first found in bone structures and then in a variety in tissues. The molecule is mainly found in regenerating tissue, as in the intestinal epithelium or in healing wounds, but also in liver fibrosis, glomerular nephritis and in various tumors. SPARC is bound, on the one hand, to collagen type I, III, IV and V. On the other hand, it can bind to thrombospondin and several growth factors within the ECM. In this way, the molecule can modulate a biological effect at the growth factor or at its receptor. A particularly interesting effect of SPARC is the ability to affect cell shape. If cells are cultivated on SPARC matrices, cell proliferation is inhibited and cell division is absent. In addition, SPARC can apparently control the quantity of the ECM formed, particularly of type I collagen.

[Search criteria: matricellular protein function]

2.4
Emergence of Tissue

2.4.1
Germ Layers and Ground Tissue

The individual functional tissues in our body develop in a long and very complex process, out of the embryonic ectoderm, mesoderm and entoderm. From there the four basic tissues are formed, which then develop into the many functional tissues, with all their characteristics, in the maturing organism.

The development of the three germ layers in humans begins in the third embryonic week (Fig. 2.47A). Up to this point, the germ disk of an embryo consists only of the ectoderm and the entoderm beneath it. Through an unknown induction mechanism, the primitive streak is formed on the surface of the ectoderm and can be recognized as a defined groove, with elevated edges, until the16th day of development. At the cranial end, the primitive streak ends at the primitive knot, which is of central importance in the further development of the germ layers. Remarkable cell changes can be observed in the area of the primitive streak. The cells round off and move into the primitive groove. This process is called invagination and has similarity with the processes that take place in the gastrulation of the amphibian embryos, which take place around the lip of the blastopore. Finally, the cells move between the ectoderm lying above and the entoderm of the germ disk below. In this way, the middle germ layer, or mesoderm, is formed (Fig. 2.47B).

The cells recently migrated into the area of the primitive knot first form a tube-like extension under the ectoderm. This is the location of the axial organ, the chorda dorsalis (Fig. 2.47C). Starting from the 17th developmental day, the mesodermal layer, as well as the chorda dorsalis separate the entoderm completely from the ectoderm. The only exception is the cranial area of the prechordal plate, from which substantial parts

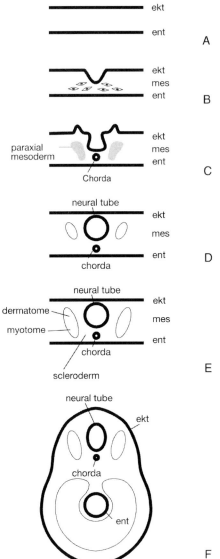

Fig. 2.47: Schematic illustration of the development of the ectoderm (ect), mesoderm (mes) and entoderm (ent) during the embryonic period. All four basic tissues develop from these germ layers.

of the head later develop. A neural tube is developed that runs through the mesenchyme (Fig. 2.47D).

The developmental phase between week s 4 and 8 week is referred to as the embryonic phase. Specific tissue and organ areas now develop from the three germ layers (Fig. 2.47E). The emergence of these areas is connected to a marked change in the outer shape of the embryo (Fig. 2.47F). At the end of week 8, the body is already recognizable in its final form. At this stage, fetal development follows until birth.

[Search criteria: embryonic development endoderm mesoderm ectoderm]

2.4.1.1 **Derivatives of the Ectoderm**

With the beginning of the third developmental week, the ectoderm of the embryo resembles another flat disk. The entire anlage of the central nervous system and the surface ectoderm stem from the ectoderm. The central nervous system anlage is called the neural plate. The formation of this structure is induced by the underlying chorda–mesoderm complex. After some days the neural plate forms two longitudinal folds with the neural groove lying between them. The folds approach each other, merge and form the neural tube, from which the complex nervous system develops. Finally, structures in the organism that provide contact to the environment develop out of the embryonic ectoderm.

The further development of the six-layered tissue in the neocortex follows a particularly structured program, which affects proliferation, cell migration and differentiation. The cortical neurons do not develop in the cortex, but rather in the more deeply positioned proliferative zone. For this reason the postmitotic neurons must pass a distance corresponding to about 500–1000 cell lengths until they reach their destination and begin terminal differentiation. The different layers are developed from the inside outward, so that the neurons must migrate through all the layers formed up to that point. The neurons must be accurately guided during this process. In addition, there are special signal mechanisms. Doublecortin (DCX) is involved, which is of crucial importance for the migration of the neurons in the cortex. The importance of DCX is to be recognized in the fact that the protein is particularly strongly expressed in growing neural tissue, while it cannot be shown in other growing tissues. Changes in the human DCX gene lead to the paralysis of brain matter, with subsequent personality changes and epilepsy. DCX is a 40-kDa protein that can be phosphorylated and possesses a Ca^{2+}/calmodulin kinase domain. This means that DCX, clearly, belongs to a cellular protein family able to direct neural migration via Ca^{2+} signaling.

Not only neuron migration, but also the growth of axons belongs to the development of the brain. This type of elongation growth, the correct guidance and branching of the axons is a fundamental step in the emergence neural tissue. The Rho family of GTPases, which convert extracellular signals and direct the cross-linking of the actin filament cytoskeleton, plays a substantial role. Rac is of particular importance, because it is involved in the extension and targeted branching of the axon. Through loss-of-function mutants, it can be shown that the loss of Rac1, Rac2 and Mtl activity leads first to defects in the axonal branching, then in propagation and, finally, growth is impaired. It is still not known how the differential activation of these individual steps is controlled.

The sensory epithelia of the ear, eye and nose develop near the central and peripheral nervous systems. The entire epidermis, including hair and nails, as well as subcutaneous glands, develop from the ectoderm. In addition, mammary glands, the pituitary gland (hypophysis) and tooth enamel stem from the ectoderm.

[Search criteria: embryonic development ectoderm derivatives]

2.4.1.2 **Derivatives of the Mesoderm**

The mesoderm initially consists of a thin layer of cells between the ectoderm and the entoderm beneath it. Beginning on the 17th day of development, the number of the mesodermal cells largely increases, developing the paraxial mesoderm. In the lateral range of the embryo, this layer remains comparably thin and forms the lateral plates. Toward the end of week 3 the paraxial mesoderm is divided into individual segments, which are from now on called somites. These determine the later shape of the body. In week 4, however, the somites dissemble again. A more axial sclerotome and a lateral dermatome develop. The cells contained in the sclerotome form a loose cell network called embryonic connective tissue, or mesenchyme. The cells of the mesenchyme can now develop into fibroblasts and form an ECM out of reticular, collagenous or elastic fibers. In addition, the mesenchymal cells can differentiate into chondroblasts and osteoblasts, which then migrate to the chorda dorsalis and build the pre-formation of the spinal column.

Aside from hyaline cartilage, humans contain fibrous and elastic cartilages. The different types of cartilage are frequently lumped together for the sake of simplicity. It must be expressly pointed out, however, that all three kinds of cartilage are completely different tissues, occurring under different circumstances, and with different composition and functions. In addition, it should be considered that all replacement bone of the human skeleton, with the exception of some dermal bones in the skull, are first composed of hyaline cartilage. The majority of these cartilage formations are then converted into bone tissue, up until adulthood. A variably thick deposit of hyaline cartilage only remains on the respective joint surfaces. Elastic cartilage in the ear lobe, the larynx and on the nose has completely different places of origin. Fibrous cartilage in the intervertebral disks develops through unknown mechanisms in the segmenting phase, between two developing vertebrae (Fig. 2.48). So far, no experimental work has been found that shows how elastic cartilage elements or fibrous cartilage develop out of hyaline cartilage.

The cells remaining in the dermatome continue to develop into the myotome. The musculature of the corresponding body segment develops from this. Histologically, the resulting myoblasts can be clearly identified on the basis of their pale nuclei. Skeletal muscle fibers result from the fusion of chains of myoblasts. With the fusion, a

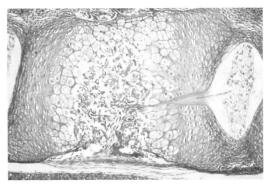

Fig. 2.48: Microscopic illustration of vertebra formation. A preliminary stage first develops out of hyaline cartilage, which is then replaced through endochondral ossification.

syncytium develops, containing many nuclei. A further cell population of the dermatome builds the dermis, as well as subcutaneous fat and connective tissue.

The precursor material of the urinary organs develops from the intermediate mesoderm. This area is called the nephrotome. From this, the first primitive kidneys develop, then the middle kidneys and finally the definitive kidneys. The importance of this developmental stage is that the primitive and middle kidneys degenerate, and only the definitive kidneys remain. Connective tissue and the musculature of the trunk wall, as well as the ribs, develop from the parietal mesoderm. The connective and muscle tissue of the gastrointestinal tract, in contrast, develop from the visceral mesoderm. In a similar manner, the mesothelial cell layers, such as the peritoneum, pleura, pericardium and epicardium, develop out of this tissue.

Starting from the third embryonic week, the first blood vessels, in the form of blood islands, develop in the mesoderm. Angioblasts, and later endothelial cells and blood progenitor cells, are to be found in the blood islands. Connected blood vessels result from outgrowths of the angioblasts. In same way, the blood vessels develop that form the tubular heart.

Blood vessels must grow into all tissue except epithelia and cartilage, so that an even food and oxygen supply is ensured. In addition, coordinated mechanisms must be in process that steer the growth and branching of the vasculature. At the same time, however, an excessive vascular formation is prevented. Angiogenic characteristics have a whole set of factors, including PDGF, VEGF, IL-8 and the acidic FGF.

The development of vascular structures occurs in close interaction with the VEGF receptor, ephrin and the ephrin receptor, as well as the proteins from the angiopoetin group and their Tie receptors (tyrosine kinases with Ig and EGF homology domains). It becomes evident that VEGF only partly represents a hierarchically superordinate modulator of the complex processes in blood vessel emergence. Given insufficient angiopoetin-1, for example, or if the binding at its receptor Tie-2 is disturbed, the regeneration of new blood vessels is missing. Overexpression of angiopoetin-2, on the other hand, leads to the destruction of the blood vessels in the embryo. Other findings have also shown that Tie-2 plays a substantial role in tumors, as its extracellular domains inhibit the emergence of blood vessels.

Stimulation of the extension of the vascular system, initiated by these factors, can be inhibited with pigment epithelium derived factor (PEDF). A typical example of the development of too many blood vessels is diabetic retinopathy, caused by oxygen deficiency. Too many blood vessels destroy the light-sensitive retina in the eye. The newest data show that PEDF does not exercise its natural inhibition on vessel growth under insufficient oxygen supply. Under oxygen deficiency, PEDF apparently causes the initiation of apoptosis mechanisms in developing endothelial cells, which again causes the formation of new blood vessels.

Thus, multiple different structures develop out of the mesoderm, such as connective tissue, cartilage and bone, skeletal and smooth musculature, and cells of the blood and lymph. In addition, the wall of the heart, the blood and lymphatic vessels, the kidneys and gonads with their ducts, the spleen, and the adrenal cortex develop from the mesoderm.

[Search criteria: development mesoderm derivatives]

2.4.1.3 Derivatives of the Entoderm

The entire gastrointestinal tract as well as the esophagus, stomach, small and large intestines develop from the entoderm. In addition, the epithelial lining of the respiratory tract, as well as the parenchyma of the tonsils, thyroid gland, parathyroid glands, liver, pancreas and thymus, also develops from the entoderm. Part of the epithelial structures of the kidney, like the collecting duct system, as well as the lining of the bladder and urethra, are also formed by cell derivatives of the entoderm.

[Search criteria: embryonic development endoderm derivatives]

2.4.2
Individual Cells, Social Interactions and Functional Tissue Development

Forerunners of tissue cells, under the effect of different morphogenic factors, develop from derivatives of the germ layers (Fig. 2.49). There are a wide variety of publications covering this topic. There is surprisingly little information, however, about the following steps when functional tissues with very special functions develop out of embryonic precursors. The main questions revolve around controlling development of the polar differentiation of epithelia or why some epithelia are very tightly sealed, while others are not. Many additional questions about developmental processes remain to be resolved, such as how the correct connections between outer dendrites and axons are made in neural networks or how three-dimensional tissue networks with mechanical and functional coupling are developed and vascularized in heart muscle.

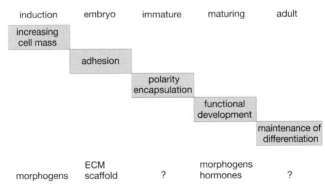

Fig. 2.49: Schematic illustration of the complex process of functional tissue development. Included are the embryonic, fetal, perinatal, juvenile and adult developmental periods. In addition, tissue development is directed on multiple cell biological levels and not only by an individual morphogen. The necessary cell mass, adhesion, polarity and functional maturation are regulated in a special temporal sequence.

[Search criteria: cell polarization functional development deviation]

2.4.2.1 Differentiation from Individual Cells

The development from embryonic to adult cells is frequently presented using the model of hematopoiesis. The cells occurring in the blood as single cells descend from hematopoietic stem cells, which occur in the stroma and fatty tissue of the bone marrow, and renew themselves throughout life through cell division. Asymmetrical cell divisions are initiated through the effect of morphogenic factors, cytokines and growth factors, in addition to the symmetrical divisions from which daughter cells develop. This serves both self-maintenance, as well as producing cells for further development. In this way, it is ensured that one part remains a stem cell in cell division, while another part develops into differentiated blood cells, along the myeloid and lymphoid pathways (Fig. 2.50). In certain developmental stages, a set of symmetrical divisions can again occur, which exclusively serve the proliferation of certain progenitor cells.

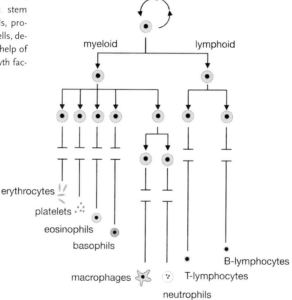

Fig. 2.50: Schematic illustration of the development of hematopoietic stem cells. Myeloid and lymphoid cells, progenitors of the individual blood cells, develop out of stem cells, with the help of morphogens, cytokines and growth factors.

In this way, for example, mature erythrocytes develop from pro-erythroblasts in several intermediate steps (Fig. 2.51). This process is not carried out automatically, but rather released by the maturing hormone erythropoetin and can be controlled by need. In order to adjust to the smaller oxygen content at high altitudes, more erythrocytes are formed than at sea level. The cells that emerge from this process as erythrocytes no longer divide and are present in isolated form for the duration of their lifespan of 120 days. Typical for this developmental pathway is the fact that from one embryonic precursor cell, individual differentiated cells develop under the effect of one individual morphogen with support of cytokines and growth factors.

In the meantime, more than 40 recombinant cytokines and growth factors are known to influence cell development in the hematopoietic system and which cause

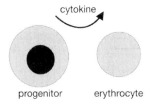

Fig. 2.51: From the progenitor to a functional individual, e.g. how a nucleus-free erythrocyte results from a progenitor cell through the effect of a cytokine.

effects via cellular mechanisms such as c-fms/macrophage colony stimulating factor (M-CSF) and c-kit/stem cell factor (SCF). This can be shown in cell culture experiments with progenitor cells, after the addition of individual cytokines. Typical of all differentiated blood cells is that they no longer divide and are broken down after a given lifespan.

[Search criteria: stem cells progenitor differentiation cytokines growth factors]

2.4.2.2 Functional Exceptions

A clear example of an exception of specific functions through the differentiation of an individual cell is the maturing a lymphocyte into a plasma cell. Here, specific antibody production begins in response to an antigen stimulus.

Under culture conditions, its move into production can be easily demonstrated. If the individual cell begins its production, this is referred to as "gain of function" (Fig. 2.52). In this special case it describes the strong regulation of an individual gene product. By definition, this process can be regarded as the simplest form of cell differentiation.

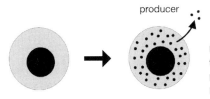

Fig. 2.52: From the progenitor to terminal differentiation. Establishment of antibody production as the simplest form of differentiation in an individual cell. This process is referred to as "gain of function".

[Search criteria: cell functional deviation]

2.4.2.3 Individual Cells and Social Interactions

Classical thought about tissue is that socially active networks of cells, fulfilling various tasks, form out of individual cells connecting to the ECM. In this sense, it is important to differentiate between the function of the individual cell and that of the actual tissue. The same applies to the development of cells and tissue. Both structures show completely independent developmental pathways, whereby embryonic cells develop into tissues with very specific functions, through different developmental steps. Thus, communicating cellular networks, with a special ECM, develop. In comparison to cell differentiation, development of specific tissue characteristics is a substantially more complex process, by which many physiological and biochemical characteris-

Fig. 2.53: Schematic illustration of the complex effect of factors during tissue emergence. Not just one growth factor is involved in cell differentiation during tissue emergence. Instead, an interaction between the ECM, cell–cell contacts and various environmental stimuli are involved.

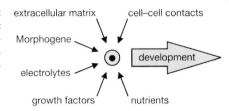

tics, apart from a definite change of shape, are changed at the same time in close cooperation with neighboring cells and the extracellular environment.

Control over the differentiation of individual cells in the hematopoietic system is one of the special tasks of the blood. The physiological development of tissue is carried out completely differently. Different cell biological regulatory mechanisms are involved. A variety of external factors, rather than an individual morphogen, influence the development in a complex cell network. These processes are not only important for the emergence of a tissue from embryonic cells, but also for its life-long maintenance. This group includes the development of cell–cell contacts, the interaction of the cells with the ECM, the nutritional and oxygen supply, as well as mechanical and rheologic loads (Fig. 2.53). Surprisingly, there is relatively little knowledge about the physiology of the developmental processes during functional tissue emergence, with the exception of bone healing in humans.

In the human embryo, the first primitive tissue of the trophoblasts and the embryoblasts emerges following the blastula stage. Preliminary stages of tissue from the ectodermal (skin, neural structures), entodermal (digestive tract, lung, liver) and mesenchymal germ layers (heart, blood vessels, connective tissue, kidney) are only developed by induction in the embryoblast. A typical example of an induction factor that stimulates development is the vegetizing factor, isolated by Heinz Tiedemann, now referred to as activin, and is able to induce the formation of mesodermal or entodermal tissue precursors. This allows the orchestration of the early embryonic development in a wide spectrum of homeoboxes.

In principle, precursors for the developing functional tissue result from the effect of a morphogen, such as sonic hedgehog or bone morphogenic protein, and thus via an induction stimulus, which begins the emergence of tissue progenitor cells from embryonic stages. Subsequently, cell movements and interactions are initiated, which can be first recognized morphologically by aggregation and later by characteristic tissue pattern formation. Thus, the direction of tissue development is first set. Now, the natural question remains how a tissue with its specific functions develops out of the still wholly immature progenitors.

[Search criteria: deviation tissue development interaction influence]

2.4.2.4 Formation of tissue

Epithelial tissue
Conditional for the development of a functional barrier is a confluent monolayer in which no gaps exist between cell connections, and where an adhesion complex exists at

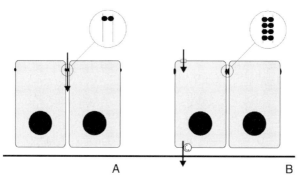

Fig. 2.54: Development of epithelial transport characteristics on a filter. After cells adhere to a filter, polarization and tight junctions are developed, with the physiologically sealing zonula occludins. If the zonula occludins, in an immature state, contains only one strand of sealing proteins (occludins), for example, then dissolved molecules may pass in a paracellularly and uncontrolled manner between the cells (A). This condition of the epithelium is referred to as "leaky." If the zonula occludins, however, matures to four to seven strands, then it seals physiologically and forms a tight monolayer (B). In this state alone the cells decide which substance is taken up by the luminal plasma membrane and secreted again by the basolateral plasma membrane, for example. Highly specific transcellular transport is developed in this way.

the border between the luminal and lateral plasma membrane. This consists of a desmosome, a zonula adhaerens and a zonula occludins (tight junction). The zonula occludins is similar to a belt consisting of four to seven interwoven protein strands. Immunohistochemically, proteins such as ZO1 and occludins can be shown here. The zonula occludins alone has the ability to form a functional barrier between the luminal and basal compartments. One can show that a zonula occludins with two to three strands does not form a physiologically intact barrier, while four to seven strands show a clear barrier, based on physiological data and freeze-fracture replicas.

In polarized cellular networks, two transport routes principally exist whereby the direction of individual transport tasks within the cell can be in opposition to one another. Some epithelia possess a paracellular (Fig. 2.54A) and some a highly specific transcellular pathway (Fig. 2.54B). Tight junctions or the amount of their anastomosing strands are decisive for a transport route. Normally, tight junctions are located at the apical–lateral cell border. Sertoli cells in the male germ epithelium, which build tight junctions at the lateral–basolateral border, are an exception. The more strands that are developed, the better the functional seal. Specially developed membrane structures in the form of channels and pumps cause transcellular transport.

At the same time that a physiological barrier develops, conditions for the polarization of the epithelium develop. In a developing epithelial cell, with still undeveloped tight junctions, proteins are built into the plasma membrane and can arrive at all points of the cell surface from there (Fig. 2.55A). However, if a tight junction is developed, proteins of the luminal membrane can no longer move into the lateral or basolateral plasma membrane (Fig. 2.55B and C). The same naturally applies to proteins

Fig. 2.55: Developmental processes in the plasma membrane of epithelial cells during polarization and functional sealing. If epithelial cells are pipetted as single cells onto a membrane, the membrane proteins are evenly distributed on the cell surface (A). In order to polarize, the cells must anchor themselves to the membrane (B). In this stage the membrane proteins are still evenly distributed. If the polarization includes the development of tight junctions, then a compartmentalization of the plasma membrane takes place, forming luminal (apical) and basolateral compartments (C). A sealing epithetlium, in principle, shows clear polarization and compartmentalization of the plasma membranes (D).

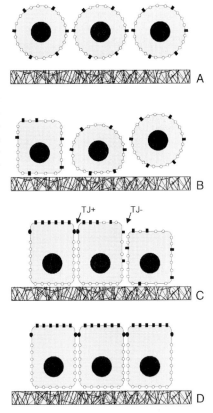

of the basolateral compartment, which can also no longer arrive in the luminal plasma membrane due to the tight junction (Fig. 2.55D).

Control over the separation of the apical and basal compartments lies in the inner leaflet of the plasma membrane and the cytoskeleton around the tight junction. This means that through the development of a functional barrier, not only close contact to the adjoining cell is achieved, but both the luminal and basal cell compartments are defined. Thus, the epithelial cell is functionally polarized. From now on synthesized proteins, such as ion channels or transporters, must not be simply sent by the cell into the plasma membrane, but particularly into either the luminal or basolateral compartments. For this purpose, "pre" or "pro" forms of plasma membrane proteins contain special signaling sequences so that they are securely sent towards and built into the luminal or basolateral plasma membrane from their place of synthesis.

For epithelial cells, it is known that a precursor of the leaf-like basement membrane is formed after an induction stimulus through secretion of ECM proteins in cooperation with neighboring connective tissue cells (Fig. 2.56A). In this way, the epithelia and connective tissue are compartmentalized. At the same time, many divisions of the epithelium are to be observed at the basement membrane (Fig. 2.56B). Consequently, the entire surface of the basement membrane can be settled, while a close

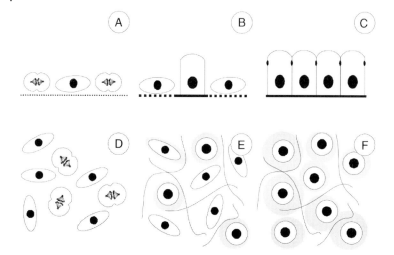

Fig. 2.56: Schematic illustration of the different pathways in tissue development. With the emergence of epithelia (A–C) and connective tissues (D–F), different developmental phases can be distinguished. Belonging to these are cell proliferation (A and D), the development of a tissue-specific ECM (B and E) and the definition of the relationship to other homogenous tissue cells (C and F).

connection to the adjoining cells is made at the same time. The polarization in the epithelium is fixed through creating close relationships by the development of cell–cell connection. This is defined by the final anchorage of the cells on the basement membrane and the development of polar characteristics (Fig. 2.56C).

With stratified epithelia, as in the epidermis of the skin, the development of functional sealing is developed in a more complicated way than with single layered, or simple, epithelia. Here, only the stratum granulosum contains structures similar to those specifically found in the tight junction (zonula occludins) of simple epithelia. Occludins (i and 10), claudines (1, 4, 7, 8, 11, 12 and 17) as well as TJ-plaque proteins, such as ZO1 can be shown immunohistochemically. How the development of the barrier function is physiologically controlled in this specialized cell layer is unknown.

Connective tissue

Connective tissue cells also divide after an appropriate induction stimulus and form cell nests (Fig. 2.56D), whereby the individual cells migrate towards one another. In contrast to epithelia, they do not build broad lateral cell contacts to their neighbors, but rather remain a discrete distance from each other. Their only contact exists over long cell runners and communication via gap junctions. The connective tissue found in development must later take on a certain size and, for this reason, a certain number of cells are needed. Here, also, the cells remain a certain distance from one another. At the same time, the intercellular spaces are built with different quantities of matrix proteins, such as fibronectin, collagens and proteoglycans, depending on the type

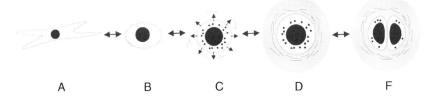

Fig. 2.57: Schematic illustration of the development of a mesenchymal cell into a chondrocyte. This development includes the transformation of a polymorphic mesenchymal cell (A) to a rounded chodroblast (B), which outlines its own lacuna, in interaction with the ECM (C). It must be determined how many chondrocytes will live in this lacuna and how the chondral capsule should be developed. According to these cell biological conditions, the mechanically loadable intercellular substance is then developed (D – E).

of tissue (Fig. 2.56E). This means that the individual cells or cell groups are still further isolated from each other through the synthesis of ECM proteins. In the last stage, cartilage cells round off and build lacunae, with specific a cartilage cap, typical for this tissue (Fig. 2.56F). At the same time, the number of chondrocytes living in the lacunae is defined.

In reality, the development of cartilage tissue is much more complex (Fig. 2.57). Its development entails, first, the transformation of a polymorphic mesenchyme cell into a rounded-off chodroblast, which designs the interaction of the lacunae with the ECM. Necessarily specified are how many chondrocytes live in this lacuna and how the cartilage cap is to be developed. According to these cell biological prerequisites, the mechanically loadable intercellular substance is then developed. A retrogression of this cell type is not intended in the natural development of the organism. At most, modifications to cells or ECM may be caused through a change in the inflammation parameters, through degenerative or inflammatory conditions, as with arthrosis or rheumatoid diseases.

Muscle tissue
The majority of our current understanding about the emergence of muscle tissue has been obtained through mouse experimentation. The emergence of skeletal muscle can be divided into different phases. Thus, during determination, it is first specified that myoblasts will be developed out of precursors. The myoblasts proliferate and move into the periphery, whereby among other things the later propagation of the developing muscle is determined. One finds dividing myoblasts after the area of propagation is fixed, whereas division is stopped in other areas. The myoblasts move closely together and fuse, thereby forming the actual muscle fiber. In this way a syncytium is developed, which consists of a common cytoplasm and multiple nuclei. Beginning with this terminal differentiation, sarcoplasm (cytoplasm), sarcoplasmic reticulum (smooth ER), sarcosomes (mitochondria) and sarcolemma (plasmalemma) of the muscle tissue are defined. A normal fiber contains 40 – 100 nuclei in a length of about 1 mm. The nuclei are oval, around 10 μm long and 2 μm thick. Due to the substantial development of the contractile elements, the nuclei are pushed to the edge of muscle fiber. Only about 3 % of the nuclei are in the fiber center. About 1 % of all nuclei must

be assigned to satellite cells. These cells lie between the muscle cell surface and the basement membrane, and do not exhibit myofibrils. They are still able to undergo cell division, and are therefore involved in regeneration and growth processes of the muscle tissue. However, the muscle tissue reacts to increased demand primarily by hypertrophy, whereby the fiber thickness increases through proliferation of the myofibrils and not by cell division. Histologically striated skeletal musculature, striated heart musculature and smooth musculature are distinguished from one another. The emergence of muscle tissue is stimulated by muscle regularization factors (MRF), including MyoD, Myf5 and myogenin. Additional effectors are different muscle enhancer binding factors (MEF). Apparently these factors work optimally when differentiation begins in the G_1 phase of the cell and the cell cycle is suspended. Cell culture experiments show that inhibitors of the cyclin-Cdk protein kinase can initiate muscle differentiation.

Nervous system tissue

The formation of nervous tissue proceeds completely differently in comparison to epithelia, connective tissue and muscle tissue. The later differentiation in this case not only depends on correct migration and differentiation of the neurons, but must additionally make specific connections to the peripheral target tissue. Axons of motor neurons must cover comparably enormous distances before innervation is complete, e.g. the distance from the spine to the sole of the foot is around 1 m. How the axon covers this distance (pathway selection) and with which molecular-biological mechanism the target tissue is reached (target selection) is largely unknown. These two developmental steps still proceed independently of neural activity. Finally, the axon must be coupled functionally to the target tissue (address selection). The neuron, preferably, receives navigational support from the ECM protein laminin, which is observed at different points on glial cells and which can signal a growth pathway. Axons of retina cells, for example, can find their target this way. Therefore, a distance of about 15 cm must be overcome from the eye background to the appropriate cerebral area. Additional adhesion molecules, such as NCAM, L1 or NrCAM, are switched on during this process. Axons can also be inhibited from a certain growth direction, and thereby are controllable through adhesion and repulsion. Changes of direction in the growth behavior are controlled by proteins from the ephrine, semaphorine and netrine groups.

[Search criteria: development tissue organ development organogenesis]

2.4.2.5 **Individual Cell Cycles**

Depending on the type of tissue, the differentiated cells are in the interphase or G_1 phase for varying lengths of time. This can persist life-long, with nerve or heart muscle tissue, and months or years, with adrenal, liver or kidney cells. With permanently regenerating tissue, e.g. the skin, interphase stop for only days. However, at this point, it is still not well known over which mechanism the individual tissue cells are directed, as they remain for such varied times in the G_1 phase. In cultivated cells, it can be shown that after addition of fetal calf serum (FCS), application of growth factors and a change in their electrolyte balance, cells in interphase can be transferred into The S phase and thus into preparation for mitosis. The cell copies its genetic material, grows and finally passes the doubled DNA on to two new cells. This process then begins again after a fixed interphase period.

A substantial question is molecular-biological regulation when cells of the G_1 arrive in S phase of the cell cycle. Normally, this step is blocked by protein Sic1. Sic1 inhibits a protein complex to which kinases such as Cdk1 belong. As long as these kinases are inhibited, the cells cannot arrive at the S phase. If Sic1 is phosphorylated several times in individual steps, however, the path into the S phase is open. An SCF protein complex along with ubiquitin complexes onto Sic1, whereby Sic1 breaks down in the proteasome and S phase can begin.

[Search criteria: cell cycle control interphase detention G_0]

2.4.2.6 Coordinated Growth

Growth of tissues can described as the increase in cell number, cell mass, the ECM and the fluid content. In development, one differentiates physiologically between fetal and postnatal growth of organs and tissues. Insulin (IN) and the insulin-like growth factor (IGF1 and IGF2) system generally work in a stimulatory manner toward growth. These hormones bind at least four receptors (INSR, IGF1R, IGF2R and Receptor X). IGF1 in the fetus still works up to the postnatal phase, while IGF2 works both in the fetus and in the placenta. It is assumed that about 45 different genes are involved in the growth of tissue. In addition to the IN/IGF, Rasgrf, Peg3, Mest and SrnpnIC are also in this category. Experiments with knock-out animals have shown that different tissues and organs are also subject to different growth control. If IGF2 cannot work because of a missing receptor, then muscle, heart, kidney and lung show minimum growth, while liver and intestine develop normally. This indicates that, aside from the superordinate IN/IGF, paracrine mechanisms for growth in the individual tissues exist about which little is known.

[Search criteria: organ development coordination growth]

2.4.2.7 Competence

There is a window of time in the development of tissues, which is only open for a certain duration and then closed again. Only during this window can certain morphogenic factors affect the development of tissue, whereas no further influence is possible afterwards. The period of the developmental-physiological reactivity of the tissue is called competence. The term originates from early embryonic development. The competence of a tissue can be demonstrated very well at the late blastula stage of an amphibian germ. At this time, the upper (animal) pole consists of competent ectoderm, which is infiltrated with the further development of the tissue of the blastopore lip. This tissue interaction is arranged to allow the ectoderm to build the neural plate (nervous tissue). From the underlying tissue, the later spinal column with the segmented muscle areas (somites) develops.

The competent ectoderm of a late amphibian blastula can be isolated and kept in culture. After application of a morphogen, such as activin, it is clear that not only neural, but also endodermal or mesodermal, tissue can develop from it. This means that the competent ectoderm shows a much larger developmental potency with *in vitro* conditions than with the development in the germ. Here, only the neural plate and the resulting nervous tissue, but no mesodermal (muscle, kidney) or endodermal (intestine) derivatives, are formed.

It would be ideal to test new morphogenic factors if permanently competent ectoderm were available. Unfortunately, this is not the case. For use in tests, the competent ectoderm must be isolated from late blastula stage quite laboriously under a preparation microscope with platinum wire loops and then cultured. If one tests the respective factors in freshly isolated tissue, appropriate tissue development can be shown after some time in culture. Isolated ectoderm behaves completely differently, if allowed to grow in culture conditions for approximately 5–6 h. If one tests activin on the artificially aged competent ectoderm, no further effect can be shown. However, if one allows the isolated tissue to grow between 1 and 6 h, then increasingly less morphogenic reactivity is present over time. In this way it is possible to observe the clearly time-dependent loss of tissue competence. This competence is, therefore, temporally limited and can be initiated by the factor Pax6 in completely different tissue progenitor cells for a time window of only few hours.

[Search criteria: tissue competence development induction]

2.4.2.8 Morphogenic Factors

Morphogenic information is transferred over paracrine factors, ECM and cellular contacts (Fig. 2.58). During the competence time, the respective progenitor cells must receive information about tissue development as fast as possible in order for the process to be introduced at all. This can take place over instructive or permissive interactions between neighboring cell populations. In this way, the epidermis of a bird, through the secretion of morphogenic factors, such as sonic hedgehog and TGFβ2, signals that flight feathers must be developed on the wings; in other areas, cover feathers or claws. A similar development mechanism is to be assumed

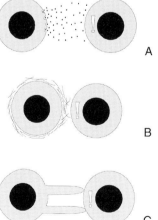

Fig. 2.58: Illustration of the interaction between tissue cells. During tissue development, morphogenic information about paracrine factors (A), the ECM (B) and cellular contacts (C) is transferred.

for the development of human skin, hair and nails, as well as the sebaceous and sweat glands.

The cellular interactions mediated over paracrine mechanisms and growth factors [growth and differentiation factors (GDFs)] have particular importance during tissue development. Four protein families belong to this category: the FGF, hedgehog (hh), wingless (Wnt) and TGFβ superfamilies.

FGFs contain a dozen structurally similar molecules, which can occur as hundreds of protein isoforms through RNA splicing. FGFs bind FGF receptors (FGFRs). On the cell exterior, the respective FGF molecule binds to the receptor. Inside the cell is a resting tyrosine kinase which is activated by the connection and consequently phosphorylates a neighboring protein, transferring it into the biologically active form. Thus, completely different developmental and functional mechanisms in the cell can, in turn, be released. Examples are the new development of blood vessels, the formation of mesenchyme or growth of axons in neural tissue.

Belonging to the group of hedgehog proteins (sonic-shh, desert-dhh, indian-ihh) are paracrine factors, which are able to form special cell types and which create natural borders between different tissues in the embryo. Sonic hedgehog, for example, has a substantial influence on the emergence of the spine. On the one hand, it structures the entire neural tube and ensures that, in the course of development, motor neurons come to lie ventrally and sensory neurons, dorsally. Development of the axons is affected by neurolin and reggie1. On the other hand, sonic hedgehog steers the segmental development of the somites and the chondralization of the spine. Desert and indian hedgehog, in contrast, steer bone growth and sperm cell formation for a long time after birth.

The Wnt family consists of cysteine-rich glycoproteins. While sonic hedgehog mainly directs the ventral tissue development in the organism, such as chondralization of the spine, Wnt1 has an influence on the more dorsal lying cells, causing them to form the necessary musculature. Wnts have substantial control functions in the emergence of the extremities and the urogenital system.

The TGFβ superfamily includes proteins such as activin, bone morphogenic proteins (BMPs) and the glial-derived neurotrophic factor (GDNF). It is well known that these proteins substantially affect the formation of the ECM proteins. This takes place both over an increased collagen and fibronectin synthesis, and over the inhibition of the matrix dismantling. TGFβs control the growth of epithelial structures in the kidneys, lungs and salivary glands. The BMPs influence n completely different cellular processes, like cell division, apoptosis, migration and differentiation. Apart from cartilage and bone development, they are involved during the polarization of the spinal cord and eye development.

[Search criteria: morphogenic factor growth FGF BMP TGF]

2.4.2.9 Apoptosis

Without apoptosis, the hands and feet would not be clearly defined extremities, but only awkward cell heaps. In the embryo, the extremities are still recognizably unstructured cell masses. The individual fingers and toes are not yet separated from each

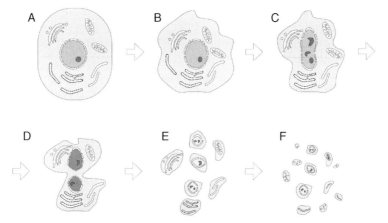

Fig. 2.59: Illustration of the different stages of apoptosis. (A) Intact cell. (B and C) Condensation of the chromatin and beginning of cell shrinkage. (D–F) Fragmentation of the cell into apoptotic bodies.

other. The gaps result from apoptosis. Similar modulation processes are to be observed with the contours of the face or internal organs, e.g. the kidney or liver. Thus, apoptosis performs an important function in the equilibrium between proliferation, differentiation and the breakdown of cells.

Apoptosis is initiated by a suicide signal and leads to the elimination of cells within hours (Fig. 2.59). Genes such as ced-9 and ced-3 are involved. CD95 and APO-1, which pass on the signal into the inner cell where the suicide program is started, are to be found on a cell death receptor. This process, however, is subject to the control of p53 and proteins of the Bcl family.

At the beginning of apoptosis, the membrane potential in the mitochondria breaks down. This again causes the release of certain molecules into the cytoplasm of the cell. One of these key molecules is cytochrome c, whose release affects an irreversible caspase-dependent cell death. The release of cytochrome c is again adjusted by the pro-apoptotic or anti-apoptotic proteins of the Bcl-2 family. The rearrangement of cytochrome c causes an interruption in electron transport between the electron transport chain complexes, III and IV, which in turn prevents the formation of ATP. The cytochrome c, secreted in the cytoplasm, binds to apoptosis protease activating factor-1 (Apaf-1), which binds caspase-9 and ATP. The developed apoptosome is now able to split caspase-3, which leads to the activation of the caspase cascade.

In the embryo, apoptosis has shaping or morphologic characteristics, whereas in the tissue of the adult humans, it frequently exhibits degenerative characteristics. This is clearly seen with the emergence of the brain. Nerve cells are first produced in surplus, which are again up to 90% lost through apoptosis. The mass of nerve cells remains approximately constant into adulthood, while it decreases by about 15% around the fourth year of life.

An impressive example of temporary apoptosis is the female breast. After the nursing period is complete, milk-producing cells become useless and must be destroyed. Apoptosis is also to be found in the uterine mucosa, during menses and around 90%

of sperm cells are destroyed by apoptosis, if they do not correspond to the cell biological quality criteria any point.

[Search criteria: apoptosis programmed cell death]

2.4.2.10 Necrosis versus Apoptosis

Aside from the targeted, directed apoptosis, non-specific necrosis can occur in tissue. In this case, completely different materials are release from tissue cells. The cells swell uncontrollably and then and burst. The free remainders initiate various inflammatory processes.

Analytically, apoptosis and necrosis should be clearly distinguishable and are quickly detectable. A relatively simple indication can be accomplished with immuno-histochemical methods. Single-strand DNA (ssDNA) can be detected in apoptotic cells in sections of tissue using specific antibodies. Necrotic cells, in contrast, can be detected with the TUNEL principle. As a control, the samples with an antibody are counter-marked for the ssDNA. Necrotic cells can clearly be differentiated from apoptotic cells in this way.

[Search criteria: necrosis apoptosis cell death inflammation]

2.4.2.11 Terminal Differentiation

The mechanisms which guide terminal differentiation are to a large extent unclear. By terminal differentiation, one considers a mostly irreversible high-grade specialization of the cell. This includes developmental steps, as with osteoblasts, which collectively take part in the formation of bones, while osteocytes wall themselves in and from this time on take over the structure of the mechanically loadable bone matrix in their environment alone. Something similar applies to the chondroblasts, which produce mechanically firm chondral ground substance on a joint surface, first in the loose cell network and then later as isogenous groups of chondrocytes within a lacuna.

Terminal differentiation is also observed if epithelia, with its various barrier characteristics and selective transport functions, develops from embryonic structures. So far it is known that it is not an individual morphogen, but a variety at factors that influence terminal differentiation. These include physical-chemical factors, like mechanical load, constant oxygen supply, pH and temperature, as well as a particular electrolytic and nutritional environment for each tissue. The information sequences built in ECM proteins of the epithelial basement membrane are surely also of particular importance.

In the course of terminal differentiation, it is specified whether the developing tissue will renew itself within days, as in many parts of the digestive tract, or whether no further cell divisions will occur life-long, as with the heart muscle or neural structures. Of particular interest are processes in the epithelial tissue regeneration of the digestive tract, in which enterocytes and goblet cells in the small intestine are renewed within a few days, whereas in the directly neighboring crypts, Paneth cells or enterochromaffin cells show regeneration cycles of months. Accordingly, there must be regulatory mechanisms that, in the context of the fast regeneration, move the corresponding stem cells from one asymmetrical mitosis to the next, whereas closely neighboring cells are held in the functional phase (interphase) for long periods without the influence of proliferation.

Finally, there are mechanisms that keep most tissue cells consistently in their location. In this way, a fibrocyte, in contrast to a fibroblast, in direct proximity to the collagen fibrils of a tendon does not leave its place of origin, as a parenchymal cell of the liver does not leave the Disse area. These processes are apparently affected by interactions between cells, as well as with the ECM and the surrounding microenvironment. Examples for all tissue types and organs could be extensively listed. However, the molecular mechanisms involved in the maintenance of cell differentiation in these cases remain unknown.

[Search criteria: terminal differentiation development]

2.4.2.12 Adaptation

Tissues do not only develop – depending on the kind of the tissue they can present different changes which are partly still physiologically healthy and partly show pathological changes (Tab. 2.5). While hypertrophy and hyperplasia are based on proliferation of the living substance by an increase in the size or number of cells, a reduction in cell size takes place in atrophy. Numerical atrophy, or involution, shows a gradual reduction in cell number.

Tab. 2.5: Possible progressive and regressive tissue changes.

Progressive tissue changes	regressive tissue changes
Hypertrophy	aplasia
Hyperplasia	hypoplasia
Atrophy	involution
Differentiation	dedifferentiation
Apoptosis	degeneration
Necrosis	

Tissues with a weak tendency to regenerate can hypertrophy in adjustment to a loss in function (compensatory hypertrophy) or can show increased performance (functional hypertrophy). In addition, tissues can hypertrophy if they fill out areas that result from the degeneration of other tissues.

Those tissues which are highly differentiated and show little or no further tendency to divide tend to hypertrophy. Hyperplasia, in contrast, arises in all tissues, where many dividable cells are still present. A good example is erythropoiesis during a stay in a mountainous area. Due to the decreased oxygen content, a reactive increase in red blood cell formation results.

Atrophy is a process that progresses in the opposite direction to hypertrophy. Here, a decrease in cell volume and tissue mass is to be observed. A typical example is the atrophy of bone and muscle that develops with persisting inactivity during bed rest. With senile atrophy there is a reduction in the tissue mass of the brain, liver and skin. If recovery of the tissue is connected with a reduction in cell number, and if the regressed tissue is replaced by adipose tissue, for example, one speaks of an involution. In the thymus, this process takes place already in youth. After the breastfeeding period, an involution of the mammary parenchyma takes place.

While atrophy leads to a qualitative reduction in the tissue mass, conserving its function, the cellular structure and associated function are strongly pathologically altered in degeneration.

[Search criteria: adaptation hypertrophy]

2.4.2.13 Transdifferentiation

The transformation of a differentiated tissue into another differentiated tissue is referred to as metaplasia or transdifferentiation. The converted tissue has a close onto-genetic history to the resulting tissue. Transdifferentiation takes place with cells in interphase and is frequently found in tissues exposed to a chronic stimulus. It therefore represents an adaptation of the tissue. An impressive example is the bronchial epithelia, which can rebuild itself into squamous epithelia with and without keratinization. It can also produce an increasing transformation into mucous-building goblet cells (goblet cell metaplasia) or to prevailing basal cells (basal cell hyperplasia). Transdifferentiation is also to be observed with cartilage and bone tissue if fiber-rich connective tissue is formed. Transdifferentiation is considered to be reversibly tissue development, but can already occur be the preliminary stage of neoplastic processes.

[Search criteria: transdifferentiation development]

2.4.2.14 Multifactorial Differentiation

The cell biological mechanisms described show that during tissue emergence, a multi-factorial occurrence, including cellular and extracellular regulation at different levels, is present. Determination can be described in the same way as with the hematopoietic system where embryonic cells are stimulated by a morphogen (i.e. bone morphogenic protein) to develop into certain tissue cells. A certain quantity of determined cells is then needed. Proliferation serves to increase cell number in order to form a tissue of the necessary size. Surfaces must be covered and three-dimensional interstitial areas settled with the same cells. In the interaction phase, the respective cells must develop in connection with an ECM typical to the tissue (Fig. 2.60). This is formed only partly by the tissue cells involved and partly by neighboring, and thus varied, cell types. In the following communication phase, the tissue cells define their social needs. Epithelia, musculature or neural structures need a very close relationship with their environment, whereas with connective tissues such as bone, cartilage or fat, defined cell distances from each other and cells in their respective ECM are necessary for their later function to be developed. It is unclear at this time whether it is already specified if cells are to still divide in each tissue. The necessary nutritive factors are not to be neglected. With exception of epithelia and cartilage, the optimal supply of the tissue with blood vessels is necessary. Finally, physical/chemical influences are particularly important.

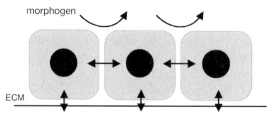

Fig. 2.60: Illustration of the interaction between cells, the ECM and morphogens during tissue development. Tissue cells in the developmental phase communicate to a considerable degree with neighboring cells and the ECM. The primary signal for tissue development is the effect of a morphogen. The resulting cell–cell interaction determines whether a close or distant spatial relationship to the neighboring cells is developed. At the same time, it is assumed that a basement membrane is developed in connection with epithelia and a very specialized pericellular matrix is developed in connection with connective tissues.

In some tissue, structures must be exposed to mechanical and rheologic stress in order to form and develop their typical function.

The list of the cell biological factors involved is still incomplete due to missing experimental data, yet they provide a glance into the different levels of control in the development of a tissue. The described influences on differentiation do not all happen in sync, but shift temporally and within certain time windows of competence. It remains unclear whether they are successively carried out or are overlapping.

[Search criteria: development determination proliferation morphogen]

2.5
Regeneration

With regeneration or wound healing, locally limited processes take place in the adult organism that largely resemble the formation of tissues during embryonic or fetal development.

2.5.1
Events Immediately after an Injury

Immediately after an injury of a tissue, blood components and diverse growth factors are found in the appropriate region. Platelets are activated whereby a clot is formed as a provisional wound closure, which consists of platelets, fibronectin, fibrin, and small quantities of tenascin, thrombospondin and SPARC. At the same time, mast cells are activated, which empty their granula and set free vasodilating, as well as chemotactic, factors by which blood cells migrate into the region. The fibronectin and fibrin clot serves as a provisional ECM for the migrated leukocytes, fibrocytes and keratinocytes. During skin injury, the keratinocytes move over the ECM, which contains vitronectin and collagen type III, in addition to fibronectin and tenascin. The moving keratino-cytes develop specific integrin patterns in order to be able to bind to the ECM. At the same time, in connection with matrix metalloproteinases, the provisional ECM is again broken down so that a new and thus tissue-specific ECM can be formed. It becomes clear that without the close interaction between cells and the ECM, no wound healing can take place.

2.5.2
Wound Closure

After a skin injury, sufficient quantities of keratinocytes must be formed so that a permanent wound closure can develop. First, a constant mitotic division of the cells in the stratum basale of the area around the wound edges is necessary. The cells formed, however, can only migrate to the wound if a suitable ECM has been devel-oped. This matrix is formed by migrated fibroblasts, monocytes, macrophages, lym-phocytes and endothelial cells. It consists of fibronectin, hyaluronic acid, and type III and I collagens. In addition, growth factors are set free from the blood platelets and the cells specified above. It is this interaction that enables the keratinocytes not only to divide in the intact stratum basale of the epidermis, but also to cover the wound. Next, laminin and collagen type VI are formed, which lead to the spreading of fibro-blasts and endothelial cells, and thus to the formation of capillaries. The releasing signal is the secretion of VEGF and FGF.

During wound healing, fibroblasts can transform themselves into myofibroblasts. This is observable in the expression of α-actin filaments. During this process, heparin inhibits the proliferation of fibroblasts and leads to the expression of α-actin filaments. It is remarkable that aside from its well-known anti-coagulating effect, heparin is able to induce this special differentiation. It is possible that growth factors, such as tumor necrosis factor-α (TNFα) and TGFβ bind to heparin, which leads to a change in the differentiation. The differentiation of myofibroblasts is again supported by the ECM. Culture experiments show that fibroblasts without mechanical load remain fibro-blasts, while the experimentally produced stretching of a collagen matrix leads to re-organization of the myofibroblasts.

2.5.3
Programmed Cell Death (Apoptosis)

In the further wound healing processes, superfluous fibroblasts, myofibroblasts and endothelial cells die again. At the same time, the remaining fibroblasts build up the interstitial collagen. Cell biologically, these processes are directed by apoptosis. The beginning of apoptosis is apparently initiated by the formation of new ECM, different growth factors and by the mechanical load.

During apoptosis, a cellular program is carried out that allows the cells to die within hours. At the same time, coordinated removal of the cell debris takes place. The chro-matin condenses, DNA fragments, the cell shrinks and protrusions appear on the cell surface, which are pinched off by blebbing. Finally, the entire cell disintegrates into fragments that are surrounded by parts of the plasma membrane. The "apoptotic bodies" are then taken up by neighboring macrophages.

2.5.4
Cooperative Renewal

Growth factors and cytokines are of great importance in wound healing. The migration of blood cells to the tissue, tissue transformation and repair mechanisms must be coordinated within a complex set of rules, so that new and therefore functional tissue can develop. A variety of biologically active substances are involved. An example of the multiple effects that growth factors have during these processes is see with PDGF. PDGF was first described as a mitogen that can stimulate cells to divide. This makes sense, as without new cells no tissue formation can take place. Additionally, however, PDGF steers cell migration and activates different cells (specifically their gene expression), something not evident before the effect of a growth factor.

At the beginning of wound healing, the blood platelets secrete the contents of their granula into the area of the injury. This process is initiated by thrombin and cofactors, by beginning the formation of a blood clot in the damaged area. First, neutrophilic granulocytes and monocytes migrate to the coagulate formed by the blood clotting, and then fibroblasts appear. Growth factors, cytokines, prostaglandins and leukotrienes are produced through this process, which in turn activate the formation of collagen, the proliferation of fibroblasts, and the formation of smooth muscle cells. The collagen scaffolding does not remain intact, but rather is repeatedly broken down and rebuilt with new three-dimensional linkages over the course of the time. The actual tissue transformation begins about 2 weeks after the formation of the wound and carries on for over a year. It is only through the close interaction between PDGF and the cells in the regenerating tissue that this process can progress without complications.

PDGF is not only present in platelets, but is found in a similar form in endothelial cells, muscle tissue, glial cells and neurons. The isoforms of PDGF (AA, AB and BB) bind to PDGF receptors. These are membrane-bound tyrosine kinases (receptor tyrosine kinases) that pass the external signals to the metabolism inside the cell. PDGF is able to activate resting cells, which then begin to divide. Intercellular communication changes which then affects migration. Small inducible genes (SIGs) and the JE gene is involved in this process. The synthesis of proteins, such as MCP-1, MCAF and SMC-CF, which possess chemotactic characteristics and can attract leukocytes or monocytes is controlled by these genes. Interestingly, glucocorticoids can inhibit the induction of JE genes and therefore affect wound healing.

Apart from PDGF, TGFβ can be identified in the granula of blood platelets. Together, the factors synergistically increase collagen production, as well as the DNA and, therefore, protein content in regenerating tissues. An individual application of PDGF can increase the emergence of granulation tissue by around 200 % and accelerate re-epithelialization, as well as the emergence of new vasculature. PDGF-BB, bFGF and EGF in combination increase re-epithelialization, whereas TGFβ alone inhibits this process. On the other hand, PDGF-BB in combination with TGFβ can cause a massive migration of fibroblasts to the wound, which results in increased synthesis of the ECM. Obviously, PDGF cannot stimulate the synthesis of pro-collagen type I alone. Only if macrophages are activated will they synthesize TGFβ, which then causes an increased synthesis of pro-collagen type I in fibroblasts. This example shows that it is not an individual factor, but usually a whole group of factors that direct regeneration processes. Frequently, these do not occur all at the same time, but rather during a certain time window, similar to the competence phase in embryonic tissue.

Apart from the migration of blood cells and the activation of fibrocytes, new blood vessels are added during tissue regeneration, for a constant oxygen and nutrient supply. Mechanically damaged endothelial cells and burn wounds set large quantities of bFGF free. This factor works, on the one hand, as a mitogen and ensures that endothelial cells divide. On the other hand, it stimulates the formation of special integrins, which affect the direction of migration, and the formation of capillary lumens. At the same time, cell–cell communication is promoted by the increased development of gap junctions in the developing tissue.

[Search criteria: regeneration wound healing growth factors]

3
Classical Culture Methods

3.1
History

A cell develops in a tissue through a variety of steps, which begin in early development and end in the adult organism with terminal differentiation. In the course of development from the oocyte to a multicellular organism, cells first form the germ layers of ectoderm, entoderm and mesoderm. Only much later do functional tissues develop, which gradually acquire specialized functions. This specialization of cell activity is closely linked to structural changes and is designated as cell differentiation. The rationale behind this development is that specialized cells in networks can fulfill their tasks much more effectively than little or undifferentiated cells in the embryonic state. Beyond that, it must be taken into consideration that some cell activity is only acquired in its entirety postnatally.

The development of an oocyte into a growing embryo already fascinated science 100 years ago. It was a time in which the biological origin of humans was being critically examined, and therefore the phylogenesis and ontogenesis of vertebrates were systematically investigated. How an organism with its various organs and differentiated tissue cells could develop from more or less dissimilar embryonic cells was discussed. In 1892, August Weismann established the theory that all organs were already mosaically fixed in the earliest developmental stages of an embryo. This assumption was a stimulating, philosophically and biomedically provocative challenge, which at that time still had no experimental basis. Only years later did H. Endres (1895) and A. Herlitzka (1897) introduce experiments in which a fertilized and developing oocyte divided into the two-cell stage. They were able to show that each of the two isolated cells developed into complete, if accordingly smaller, amphibian embryos. Thus, Weismann's theory was disproved.

[Search criteria: cell tissue culture historical review]

Tissue Engineering. Essentials for Daily Laboratory Work W. W. Minuth, R. Strehl, K. Schumacher
Copyright © 2005 WILEY-VCH Verlag GmbH & Co. KGaA, Weinheim
ISBN: 3-527-31186-6

3.2
First Cultures

There are different views about the beginnings of cell and tissue culture. The fact is, however, that working with living structures under *in vitro* conditions did not develop from one day to the next, but rather today's technical position has been achieved after almost 100 years of development. The development of cell culture happened almost in parallel with the technical use of automobiles.

The driving force for *in vitro* investigations with cells was the enormous interest in the developmental processes of the organism at the beginning of the 20th century. The ideal opportunity for the observation of such embryonic development was during the spring season. All that was necessary to make the observations was amphibian eggs, fresh spring water and a magnifying glass. No infections occurred in the fertilized eggs, as they developed a transparent, gelatinous membrane around themselves. There were no nutritional problems as amphibian eggs, similar to chicken eggs, are rich in yolk (polylecithal). This means that nutrition is built into the individual cells in the form of yolk plaques and is passed on with each division. This yolk represents a sufficient nutritional reserve until the time that the larvae emerges. Embryonic human cells, in contrast, possess no considerable yolk supply (oligolecithal/alecithal) and during their development must be supplied first via the extracellular environment, through trophoblasts and later via the placenta.

The observations of, and manual interference with, living cells became more difficult as the egg membrane of the amphibian embryos was removed. Attempts were made to isolate and further cultivate individual cells from the embryo in order to understand the developmental potency of certain germ areas. After removing the protecting egg membrane, one was suddenly confronted with the problem of infections. Sterility was still a little known concept at this time. Antibiotics were also unknown. Infectious agents introduced in into the culture could not be treated. With an infection, the isolated cells and developing tissues would become overrun with bacteria or fungi.

The internal environment of cells was basically unknown into the 1950s. It was with complete surprise that one observed how isolated amphibian germ layers swelled after removal of their gelatinous membrane in water, used as the storage medium at the time, and finally burst. The composition of the cytoplasm or the electrolytic content of an isotonic salt solution was not yet known. Through experimental work in innumerable series of tests, it was finally recognized that electrolytes like sodium, chloride, potassium and calcium were indispensable components of the physiological environment. It took decades before amino acids, nucleic acids, glucose and vitamins were added as essential components of culture medium.

Only at the beginning of the 1960s did cell culture techniques reach such a high standard that some embryonic cells could be kept in culture for relatively long periods. Under good culture conditions, cells would even develop into tissue structures. In parallel, the beginnings of modern biotechnology were developing. It was recognized that some of these cultivated cells were suitable for reproducing viruses. Accordingly, vaccines for viral infections, e.g. the cause of child paralysis (polio virus), were developed. Apart from different tumor cells, interestingly enough, kidney cells proved particularly suitable for the proliferation of viruses. Most culture media available today originated from this time, as well as many cell lines offered in catalogs by the most diverse cell banks.

3.2.1
Culture Containers

3.2.1.1 Individual Culture Containers

Today there are a variety of culture containers in different sizes, with concave, convex or flat growth surfaces. In principle, however, all of these originate from the classical Petri dish. Containers made of polystyrene have now have replaced glass containers to a large extent (Fig. 3.1). In the past, individual dishes had to be prepared through cumbersome cleaning techniques and sterilization work. Today, disposable products are used. After opening the sterile packaging, one already has the desired container ready for experiment. The choice of a culture container largely depends on the measure and in which volume cells are to be grown. The scale in the laboratory spans the cultivation of stem cells in a hanging droplet, to roller bottles, containers or bags with multiple liters of content for antibody production with hybridomas, or to the breading of virus-proliferating cells.

Since most containers offered are manufactured from polystyrene, they can nearly always be used for adherent cell cultures. Although this material does not occur in the organism, many cells have a high affinity for it. This means that cells pipetted into the container attach more or less firmly after some time in the culture dish. The result of this is that liquid can be easily aspirated or decanted in order to change the media without loss of cells.

With suspension culture, where cells float freely in the medium, however, both cells and medium must be removed together and centrifuged. Only then can the old medium be pipetted out and exchanged with new medium, as the cells remain pelleted in the centrifuge tube.

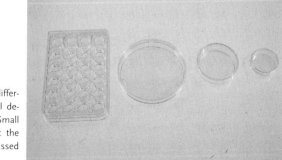

Fig. 3.1: Culture containers of different sizes and forms are used, all derived from glass Petri dishes. Small spacers in the cover ensure that the medium can be continuously gassed in an incubator.

3.2.1.2 **Dimensions of the Container**

A variety of sizes and shapes of dishes and flasks are used in cell culture techniques. Cells can be additionally held on slides for microscopy as well as in gas-permeable bags. In principle, a culture dish has a cover and a bottom part. The cover rests loosely on the bottom, held at a distance by cams that allow for gas ventilation. Flasks can be tightly closed by a screw cap. By loosening the cover a little, even ventilation can also be achieved. The growth surface, the cell density, both in seeding and in confluence, as well as the volume of the necessary medium are of importance in choosing a culture container (Tab. 3.1), depending on the type of cell and the cell line.

Tab. 3.1 Dimensions of cell culture containers. The correct container must be selected for each cell type. If maximum cell density with minimal medium is desired, dishes are the best choice. If, however, antibodies are to be harvested that are secreted into the medium, a flask is better.

		Surface (mm^2)	Cell density		Medium (ml)
			Yield (10^6)	Confluence (10^6)	
Microplates	6 well	900	0.3	1.2	4
	12 well	400	0.1	0.4	2
	24 well	200	0.5	0.2	1
Dishes	35 mm	960	0.3	1	2
	60 mm	2800	0.8	3	3
	100 mm	7800	2.2	9	10
	150 mm	17600	5.0	20	20
Flasks	T-25	2500	0.7	3	5
	T-75	7500	2.1	8	10
	T-160	16000	4.6	18	20

3.2.1.3 **Coating the Culture Dish**

Some cell types grow particularly well on a growth surface of polystyrene, while for other cell types this material is suboptimal, for unknown reasons – they do not adhere and consequently die. For such poorly adhering cells, therefore, a surface corresponding to the characteristics of the natural ECM or one in which cells can connect to via an analogous mechanism must be made available. Such coated culture dishes are commercially available, but one can also easily prepare them. Collagens, fibronectin, laminin, chondronectin or other components of the ECM can be used. According to the directions of the manufacturer, the respective protein is dissolved and spread over the entire bottom surface of the container. The mixture is then left under a sterile hood to dry overnight. The poor solubility of some of these substances can be solved by increasing the NaCl concentration in the solution buffer used or by acidifying it with HCl. It is not always necessary to coat the culture dish surface with ECM proteins. Much success has also been achieved with peptides such as polylysine.

For some years, various companies have offered culture dishes in which the bottom is physically, chemically or mechanically treated so that certain cells of a cell mixture adhere and grow particularly well, e.g. primaria dishes. The growth surface of these

- *Keratinocytes serum-free medium (SFM)* – for the culture of keratinocytes, whereby growth of fibroblasts is inhibited at the same time.
- *Knockout DMEM* – optimized for the growth of mouse stem cells.
- *StemPro* – a complete methylcellulose medium intended for the culture of progenitor cells from human hematopoetic tissue.
- *Neurobasal medium* – used for the growth of neurons of the central nervous system.
- *Hibernate medium* – used for the short-term maintenance of neural cells.
- *Endothelial SFM* – used for the growth of cow, dog and pig vascular endothelial cells.
- *Human endothelial SFM* – particularly suitable the proliferation of human venous and arterial umbilical cord endothelial cells.

[Search criteria: culture media composition]

3.2.2.1 Ingredients

Whole-cell systems have been offered for some time by different companies, and include cells from a wide range of tissues, organs and species. Media delivered with the cultures are particularly designed for the culture, with the necessary additives in order to optimally acquire more cells. The composition of the supplied media is frequently unclear. Whether it is possible to work experimentally with such unknown factors must be decided individually. For this reason it is important to have some concept of how the composition of a culture medium.

Culture media contain a variety of different components. The bases of a culture medium are the buffered salt solutions known as phosphate-buffered saline (PBS), Earle's buffered saline solution (EBSS), Gey's buffered saline solution (GBSS), Hanks' buffered saline solution (HBSS) and Puck's salt solution. The electrolytic solution that is most suitable for a particular cell type must be individually decided.

Frequently, cells must frequently be isolated from tissue before they are cultivated. It is recommended that both in the breaking down of tissue as well as with the subsequent culture work, media with the same buffered salt solutions are used in order to avoid osmolar stress.

One medium developed in 1977 is MCDB 104. It contains the following components: $CaCl_2 \cdot 2H_2O$, KCl, $MgSO_4 \cdot 4H_2O$, NaCl, NaH_2PO_4, $CuSO_4 \cdot 5H_2O$, $FeSO_4 \cdot 7H_2O$, $MnSO_4 \cdot 4H_2O$, $(NH_4)_6Mo_7O_{24} \cdot 4H_2O$, $NiCl_2 \cdot 6H_2O$, H_2SeO_3, $NaSiO_3 \cdot 5H_2O$, $SnCl_2 \cdot 2H_2O$ and $ZnSO_4 \cdot 7H_2O$. The electrolytes contained in the culture medium are necessary in order to simulate the environmental relationship within and outside of a cell, and thus to enable a cell to survive outside of the organism at all. However, it was already shown decades ago that proliferation (mitosis) of cells could be accelerated and at the same time the functional operating phase (interphase) shortened through differing compositions of electrolytes. In this way, it was possible to harvest as many cells in as in a short a time possible without increasing the amount of serum or the addition of growth factors. The additional metals and rare elements contained in the medium are needed for catalytic processes in the cell.

Tab. 3.2: The pH, electrolytes, glucose content and osmolarity in different culture media, measured in an analyzer. In no case are the electrolyte values identical with that of serum.

		Human arterial serum	IMDM	Medium 199	BME	Williams medium E	McCoy's 5A medium	DMDM
pH		7.4	7.4	7.4	7.4	7.4	7.4	7.4
Na^+	(mmol/l)	142	117	139	146	144	142	158
Cl^-	(mmol/l)	103	81	125	111	117	106	116
K^+	(mmol/l)	4	3.9	5.1	4.8	4.8	4.8	4.8
Ca^{2+}	(mmol/l)	2.5	1.1	1.5	1.4	1.4	0.5	1.3
Glucose	(mg/dl)	100	418	99	94	186	270	382
Osmolarity (mOsm)		290	250	270	286	288	289	323

If various media, such as IMDM, BME, William's medium, McCoy's 5A medium and DMEM, were to be analyzed with serum in and electrolyte analyzer and compared to the interstitial environment of an organism (Tab. 3.2), then these values would in no case agree. This touches on the fact that 30–50 years ago the goal was not to simulate the interstitial environment of tissues, but rather exclusively to optimize proliferation.

For protein metabolism, a culture medium contains amino acids such as L-alanine, L-arginine–HCl, L-asparagine · H_2O, L-aspartic acid, L-cysteine–HCl, L-glutamic acid, L-glutamine, glycine, L-histidine–HCl · H_2O, L-isoleucine, L-lysine–HCl, L-methionine, L-phenylalanine, L-proline, L-serine, L-threonine, L-tryptophan, L-tyrosine and L-valine. It is noticeable in this list that usually only the L and not the D isoforms of the amino acids are contained in the culture medium. This makes sense, since only the L form is used in protein synthesis in animal and human cells. Epithelial cells, however, exhibit a particular characteristic. They can also usually use D amino acids, since they possess an enzyme that can transform the D isoform into the L isoform. Fibroblasts cannot do this. For this reason, a medium that contains D-valine instead of only L-valine can be used to select for the growth of epithelia through the elimination of fibroblasts.

In addition, a cell in culture needs vitamins, such as biotin, choline chloride, D-Ca-pantothenate, folic acid, D,L-6,8 D-lipoic acid, nicotinamide, pyridoxine–HCl, riboflavin, i-inositol, thiamine–HCl and vitamin B_{12}. Additionally, components are needed for DNA and RNA synthesis, as well as for energy metabolism, e.g. adenine, thymidine and glucose, as well as linoleic acid, putrescine-2–HCl and sodium pyruvate.

How a media is buffered also depends on whether the culture is to be carried out in a CO_2 incubator or in room atmosphere – $NaHCO_3$ or a biologically compatible buffer, such as HEPES or Buffer All (Sigma) is used. The pH of the culture medium should be maintained at 7.2–7.4. Phenol red is added to the culture medium as a color indicator for the visual estimation of the pH. However, one should be careful if working with

[Search criteria: serum-free culture conditions growth factors media]

3.2.2.5 pH of the Medium

The maintenance of the acid/base equilibrium normally takes place using sodium bicarbonate (sodium hydrogen carbonate), which serves both as a buffering substance and as an essential nutritional component. An increase of the CO_2 content results in a decrease in the pH value, which is neutralized again by an increased content of sodium bicarbonate. Finally, equilibrium is aimed at a physiological pH between 7.2 and 7.4.

The sodium bicarbonate buffer system in the culture medium consists of:

$NaHCO_3$ and CO_2
$NaHCO_3$ dissociated: $NaHCO_3 + H_2O \Leftrightarrow Na^+ + HCO_3^- + H_2O$
$Na^+ + H_2CO_3^- + OH^- \Leftrightarrow Na^+ + H_2O + CO_2 + OH^-$

This reaction depends on the partial pressure of CO_2 in the atmosphere. Under low CO_2 partial pressure, the reaction equilibrium will move to the right, which means the medium contains more OH^- and is therefore basic. Therefore, the sodium bicarbonate buffered media are gassed with CO_2 in the incubator. The buffering effect of the system is based on the following reactions:

An increase in H^+ causes: $H^+ + HCO_3^- \Leftrightarrow H_2CO_3 \Leftrightarrow CO_2 + H_2O$
An increase in OH^- causes: $OH^- + H_2CO_3 \Leftrightarrow HCO_3^- + H_2O$

Liquid media, which are used in a CO_2 incubator, are usually held within a defined pH range through the addition of sodium bicarbonate and the controlled ventilation with CO_2 via a control valve. If these media are left for a longer time in culture dishes under a sterile hood, a pH shift into the alkaline range can be noticed through the violet discoloration. This originates from the fact that only about 0.3% CO_2 is present in the room atmosphere, whereas the incubator is usually maintained at 5% CO_2. If media must be used at room atmosphere, then HEPES, Buffer All or another biological buffer is used for the stabilization of the pH. In addition, there are culture media such as Leibowitz L15 that are equipped with a phosphate buffer system for working in the room atmosphere.

[Search criteria: cell culture medium pH more buffer bicarbonate]

3.2.2.6 Antibiotics

Antibiotics are an important aid in cell culture techniques, but should be used as sparingly as possible. On the one hand, they can damage the cells; on the other hand, their use can prevent an infection from being discovered. Individual substances such as penicillin G or streptomycin are offered, but one can also fall back on whole antibiotic cocktails, like a complete antibiotic/antimycotic solution (Tab. 3.4). A recommendation for single substances cannot be given here. Most antibiotics are supplied in a solution, ready for use. Again, aliquoting should be performed according to the quantity needed each time. Due to their possible cytotoxic effects, it should be standard practice to refrain from using antibiotics in cell experimentation. However, the

Tab. 3.4: Examples of antibiotics and antimycotics frequently used in culture media

Medication	Use	Spectrum	Stability in medium at 37 °C (days)
Streptomycin sulfate	50–100 µg/l	Gram-positive and -negative bacteria	5
Polymixin B sulfate	100 U/l	Gram-negative bacteria	5
Penicillin G	50–100 U/l	Gram-negative bacteria	3
Neostatin	100 U/ml	fungi and yeast	3
Neomycin sulfate	50 µg/l	Gram-positive and -negative bacteria	5
Kanamycin sulfate	100 µg/l	Gram-positive and -negative bacteria as well as mycoplasma	5
Gentamycin sulfate	5–50 µg/l	Gram-positive and -negative bacteria as well as mycoplasma	5
Amphotericin B (Fungizone)	0.5–3 µg/l	fungi and yeast	3
Anti-pleuro pneumonia-like organism (PPLO) agents	10–100 µg/l	mycoplasma and Gram-negative bacteria	3

use of antibiotics cannot be avoided with primary cultures where the preparation of the organism is not 100% sterile.

Contaminated replaceable cultures should be eliminated immediately. However, it is also possible to attempt to control the contamination. First, one should analyze the extent a bacterial, fungal or yeast infection with the assistance of a microbiological lab. Contaminated cultures should be separated from the non-contaminated. If, for the elimination of the infection, it is necessary to work with antibiotics and antimycotics, one should be aware that these can be toxic to cultivated cells in higher concentrations. This particularly applies to the antibiotic Tylosin and the antimycotic Fungizone.

[Search criteria: culture medium antibiotics]

3.2.2.7 Other Additives

The amino acid L-glutamine must be included in all cell culture media as an essential nutrient. L-glutamine is, however, very instable above – 10 °C so that with storage over longer periods of time it is difficult to determine how much L-glutamine is still in the media. Therefore, it is a good idea to add L-glutamine directly before using the media, according to the directions of the manufacturer. In addition, there are culture media that contain stabilized glutamine.

Hormones and growth factors are frequently needed for a SFM (Tab. 3.3). These materials have quite variable chemical characteristics. Many are extremely insoluble and must be brought into solution through a special procedure. With highly insoluble substances it helps if the additive can be dissolved in as small a volume as possible of absolute ethyl alcohol. Afterwards, the alcohol solution can be pipetted into the warm culture medium in small steps, gently swishing. It should now be noted that alcohol is present in the medium. For this reason the quantity of alcohol added should be kept as small as possible. Also, many of these additives only have a short bioavailability and should be added to the medium as shortly before use as possible. These materials are very quickly broken down in the medium and thereby inactivated. In addition, absorption by the surface of the culture container can occur, which also has an influence on the bioavailability of substances. In order to obtain precise information about the availability of additives, it is advisable to have the effective soluble content of the respective material in the medium tested once by an appropriate lab. In this way it can be easily analyzed whether the quantity of substance available really corresponds to the desired concentration.

[Search criteria: culture medium additives L-glutamine]

3.2.3
Growth Factors

3.2.3.1 Overview of Different Growth Factors

There is a unending variety of growth factors (See Chapter 2), including classical growth factors like TGFβ, VEGF, IGF, the neurotrophins, GDNF, EGF, FGF as well as PDGF. Many further factors with growth-promoting characteristics are also used, such as chemokines and interleukins. For further information, it is advisable to refer to specific literature on this topic.

- TGFβ. This factor was found for the first time in fibroblast cultures transformed with viruses. It was shown that fibroblast proliferation was uninhibited in presence of TGFβ. Almost all cells of the human body produce this factor and a particularly high concentration is found in thrombocytes. In general, TGFβ is an inhibitor of epithelial cell proliferation. The cells are blocked in the G_1 phase. There are three classes of TGFβ receptor (TGFβRI – III) as well as the endoglin receptor. The effectors of these receptors are the SMAD proteins, which have an influence on gene activity via interaction with the Ras/mitogen-activated protein (MAP) kinase signaling pathway. The individual isoforms such as TGFβ1, TGFβ2 and TGFβ3 show completely different effects on cell differentiation and proliferation. The BMPs also belong to the TGFβ family as well as the activins, inhibins and nodals, which have, among other things, a large influence on embryonic development.
- VEGF. This whole family of factors steers many developmental and growth and regeneration processes in endothelia. It consists of VEGF-A, VEGF-B, VEGF-C and VEGF-D. The angiopoetins and the ephrins also belonging to this family. VEGF affects not only endothelial cells, but also Schwann cells, and pancreas and retina cells. Receptors for VEGF are VEGFR-1 (Flt-1), VEGFR-2 (KDR/Flk-

1) and neuropilin. The signaling cascade activates protein kinase 2 (VEGFR-2) and phosphoinositol-2-kinase or MAP kinase. In endothelial cells, VEGF activates factors such as Bcl-2 and A1, which prevent apoptosis.

– IGF. This group contains IGF-I and IGF-II. Both factors stimulate growth in the entire organism. As receptors on the cell surface, they function as tyrosine kinases.

– Neurotrophins. This group contains NGF, BDNF, neurotrophin (NT)-3 and NT-4/5. These factors promote the growth, differentiation and survival of many different neurons during development and in adult tissue. With tumors such as neuroblastomas and medulloblastomas, as well as in neurodegenerative illnesses, it can be shown that these factors and their receptors, such as trkA, trkB, trkC and p75^{NGFR}, have been disturbed. The signal transduction takes place over ERK and MAP kinase.

– GDNF is related to TGFβ and was originally identified as the survival factor for dopaminergic neurons. GDNF binds to the GDNFR-α1 receptor. The signal is transmitted to the tyrosine kinase RET and the Ras/ERK pathway. Incorrect expression of GDNF is found in thyroid carcinoma and different endocrine neoplasias.

– EGF. The family also contains TGFα. Both factors resemble each other closely compete for receptors occurring in tissues all over the body, such as EGFR, HER-2 (erbB), HER-3 and HER-4. Only on hematopoetic cells are these receptors missing. Other growth factors such as heparin binding EGF-like growth factor (HB-EGF), amphiregulin and betacellulin can also be categorized in this group. The EGF signaling cascade is carried out over Ras, Raf, the MAK kinases, phosphatidylinositol-3-kinase (PI$_3$K) and PLCγ. EGF stimulates the proliferation of epithelial cells and is therefore important in wound healing. TGFα, however, varies in function depending the on the cell type and may stimulate or inhibit proliferation.

– FGF. This group contains the acidic FGF (aFGF, FGF-1) and the basic FGF (bFGF, FGF-2). Synonyms for these growth factors are endothelial growth factor, retina-derived growth factor and cartilage-derived growth factor. Under culture conditions, these factors stimulate proliferation in a variety of mesenchymal cells such as fibroblasts, chondroblasts, osteoblasts and myoblasts. FGF binds receptors such as FGFR-1 to FGFR-4, which are all tyrosine kinases.

– PDGF. Again, here are multiple factors, all of which can be found in the α-granula of thrombocytes. This group contains PDGF-A, PDGF-B, PDGF-C and PDGF-D, which can bind to different receptors, therefore causing different effects during thrombogenesis. The activated α-receptor can inhibit chemotaxis in fibroblasts and myoblasts, while the β-receptor stimulates this process.

3.2.3.2 Effect of Growth Factors

Growth factors are signaling molecules that influence cell proliferation, growth and differentiation. In most cases they are polypeptides that bind to receptors of the cell surface, thus releasing a signaling cascade, which finally affects gene expression and cell cycle activity. Growth factors are found in all tissues in which the cells divide, e.g. they are found during embryogenesis, in tissue renewal in the adult, in injuries and also in the emergence of tumors.

Most growth factors have autocrine activity. When a factor is produced and secreted by the cell, it can directly bind to a specific receptor on the cell surface again. If neighboring cells are activated by the secreted factors, the effect is described as paracrine. If a growth factor is carried through the bloodstream before reaching its target cells, then it is described as endocrine.

The effect of a growth factor is dependent on the life cycle the cell is currently in. Mitogenic, tropical and anti-mitogenic reactions can be initiated depending on the receptor and developmental stage. Only in the G_0 and G_1 phases before DNA synthesis is it possible for the growth factors to affect the cell cycle. Cell biologically, the G_1 phase can be divided into early, middle and late stages. The growth factors lead the cell through the G_1 phase up to the point of restriction, i.e. up to the end of the middle G_1 phase. Afterwards the rest of the cell cycle continues irreversibly and without any further signal from growth factors.

The growth factors bind to the extracellular domain of specific receptors in the plasma membrane. Usually these are receptor kinases, which are phosphorylated after binding of the factor. Thus, cell effectors in the cytoplasm, referred to as downstream effectors or second messengers, are activated. These molecules arrive in the nucleus and exert an influence on gene transcription, with the cooperation of transcription factors such as Fos, Myc and June. Ras and Raf proteins as well as members of the MAP kinase family are involved in the signaling cascade.

[Search criteria: cell culture growth factors addition proliferation]

3.2.4
Cell Culture Techniques

Key phrases for modern work with cell cultures are "cell culture engineering", "metabolic engineering", "bioprocessing", "genomics", "viral vaccines", "industrial cell culture", "medium design", "viral vector production", "cell line development", "process control" and "industrial cell processing". Nearly all of these are concerned with a special kind of culture. The cells in question should proliferate as fast as possible in order to synthesize a bioproduct, such as a medication or vaccine, with high efficiency. A wide variety of innovative devices have been developed in recent years for all these techniques.

Today, there are two completely different concepts in breeding cells, i.e. they may either float freely in the culture medium (non-adherent) or bind to the growth surface of a culture container (adherent). A combination of both methods consists of letting cells adhere to small porous beads and then keeping these in a constant swirling movement through agitation. In all cases part of the goal is to stimulate cells to proliferate in as simple a way as possible, in order to either use products produced by the cells or the cellular material. After deciding on a growth container for the culture, different culture

Fig. 3.2: Microscopic view of non-adherent hybridoma cells. They are roundish, non-polarized cells, which produce antibodies and grow lying on the surface of culture containers, without firmly attaching themselves.

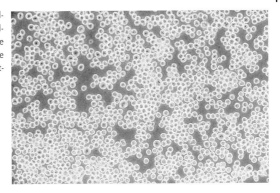

techniques and media with the necessary additives should be considered for the analytical standard. Culture should be described for a variety of different cells according to the work expenditure involved. Different cell lines are used for producing antibodies (Figs 3.2 and 3.3), e.g. Madin-Darby canine kidney (MDCK) cells as a model example for an epithelial cell line (see Fig. 3.5), epithelial cell lines in transfilter experiments (see Fig. 3.6) and isolated heart muscle cells as an example of primary cultures.

3.2.4.1 Hybridomas for the Production of Monoclonal Antibodies

Antibodies are globular proteins (immunoglobulins) formed and secreted by the B-lymphocytes of the immune system. This happens in response to the presence of a foreign substance called an antigen. Experimentally, this characteristic can be induced and maintained in hybridoma cells. In the meantime, hybridomas can be selected from the catalog of a cell bank and delivered to the lab (Fig. 3.2). Under culture conditions, the cells produce a monoclonal antibody that reacts with the desired antigen. The antibodies produced can be used for diverse immunohistochemical or biochemical detection of protein recognition, or used in Western blot.

The hybridoma cells are either sent in frozen form or as a growing cell population in a container. The following media are mainly used for their culture: DMEM, FCS 10%, Na-pyruvate 1%, L-glutamine 1% or RPMI 1640, mercaptoethanol 3 µl up to 500 ml, FCS 10%, gentamycin 1%, amphotericin 0.5%.

The goal of this type of culture is to produce as many hybridoma cells in as short a time as possible in order to gain the maximum output of antibodies. To do this, hybridomas are grown in a 24-well culture plates with 1 ml of media per well as long as it takes for the cells to cover the dish completely. At this point the cells are dense enough to be transferred into small culture flasks (Figs 3.2 and 3.3). Optimally adapted cells grow easily.

Well-growing clones can be diluted up to 20% the next time they are seeded. From the small culture flasks the cells can then be transferred step by step into the desired sized flasks. The cells are maintained in this stage for antibody production. In each case, the proliferation of special biomaterials remains as the main consideration. Understandably, the producer cells should be as simple to multiply as possible. The de-

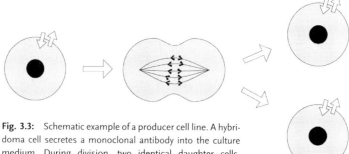

Fig. 3.3: Schematic example of a producer cell line. A hybridoma cell secretes a monoclonal antibody into the culture medium. During division, two identical daughter cells, which again produce antibodies, develop from it.

scribed cultures have the big advantage that after the initiation of the production process of a certain material, the production time, and therefore the amount of product, can be adjusted as desired, since with optimal growth conditions, hybridomas divide and produce two identical daughter cells (Fig. 3.3).

If one wants to cultivate hybridoma cells for more efficient production of antibodies, there are a large variety of coordinated and commercially available media and culture techniques available for scaling-up, which means that working in ever-larger dimensions is possible. This example clearly shows that nearly all media available up to now have been developed with the goal of enabling cells to divide as fast as possible, so that maximum synthesis is reached in as short a time as possible. This goal has frequently only been accomplished through trial and error, by adjusting the electrolyte composition and osmolarity of the respective media.

[Search criteria: hybridoma cells antibody production engineering]

3.2.4.2 Immortalized Cell Lines as Biomedical Models
Cell lines are generally differentiated into "primary" and "continuous". The cultivation of freshly isolated cells of an organ or a tissue *in vitro* is called primary culture. This will be discussed later, using the example of cardiomyocytes. If such cells grow unimpaired

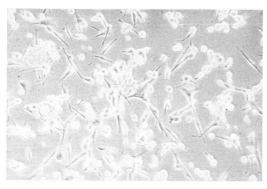

Fig. 3.4: Microscopy of 3T3 cells in culture. The fibroblast-like cells cling to the surface of culture dishes and develop three-dimensional nets.

Fig. 3.5: Microscopy of adherent epithelial cells. MDCK cells build a polar, differentiated epithelium on the surface of a culture dish after attaching. ECM proteins are thereby secreted into the culture medium without a basement membrane being developed.

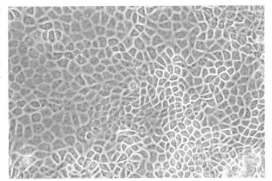

and divide, they must eventually be distributed into new culture containers. This happens when the culture dish is completely covered with cells, i.e. a confluent monolayer is formed. The cells are removed from their confluent dish and distributed in subsets into new dishes or flasks. Without subcultivation, the cells will eventually die. From this first subcultivation on, the cells are referred to as a primary culture. If a cell line can be subcultivated more than 70 times after the primary isolation without restriction, it becomes a continuous cell line, by definition. An example is the fibroblast-like 3T3 cell line (Fig. 3.4). During long-term cultivation, the characteristics of cells do not usually remain constant. Not only can typical characteristics of the original cells be lost, but atypical characteristics may be acquired.

A good example of a continuous epithelial cell line is the MDCK line (Fig. 3.5). The MDCK cell line originates from the kidney of a cocker spaniel, and was isolated and cultured in 1958 by Madin and Darbin. The 49th subculture was supplied to the American Type Culture Collection (ATCC; Manassas, VA 20108, USA). Today, this line is available in different subculture stages, and there are two varieties (strain I and II) with different morphologic and physiological characteristics. In addition, there are multiple clones with completely different characteristics stemming from each variety. The cells are stored and delivered frozen.

MDCK cells have mixed characteristics and can therefore not be clearly attributed to a particular segment of the renal tubule. The cells of the continuous line mostly differ

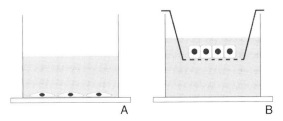

Fig. 3.6: Example of modulating a phenotype under culture conditions. Schematic representation of MDCK cells on the surface of a culture dish as a flat monolayer (A) and in a filter cartridge that supports polar differentiation (B).

from those of a primary line by an altered number of chromosomes. Like many other cell lines, MDCK cells form an epithelium in culture, but no basement membrane typical for this tissue. Instead, the cells secrete basement membrane components such as fibronectin and type IV collagen in soluble form into the culture medium. Clearly, they have lost the ability to bring these proteins to the basolateral side of the epithelia in insoluble form and to interlace them three-dimensionally into a functional basement membrane (Fig. 3.5).

Depending on the culture container and available growth surface, MDCK cells exhibit completely different characteristics. In polystyrene culture dishes, the cells grow as a monolayer, in a single flat cell layer (Fig. 3.6A). Either spontaneously or after application of hormones, the cells build hemicysts, referred to as domes and blisters. MDCK cells adhere reversibly to the dish surface, and can be detached by means of trypsin and EDTA for subcultivation. On special filter systems, the MDCK cells develop into a more strongly polarized epithelium with physiological transport characteristics (Fig. 3.6B). Although the MDCK cells otherwise behave in their proliferation like tumor cells, they have a limited lifespan in this differentiated condition and subculture is often no longer possible. Master cultures are, therefore, always cultivated in plastic containers rather than on filters. Depending on the investigation conditions, cells are taken out of the master cultures and cultivated accordingly. Frequently, one uses MDCK cells as a host for the proliferation of a virus or as an epithelial cell model for investigating the molecular mechanisms of transport processes.

With the culture of MDCK cells in a Petri dish, the apical and basolateral sides are in contact with the same medium. This is not the case with polarized epithelial cells in the organism, since completely different environmental conditions always prevail apically and basolaterally. Therefore, a biological short-circuit frequently develops in the culture dish, through equal conditions on the luminal and sides, which can inhibit differentiation of the cultures. If the cells develop physiological seals at a later time, such as with tight junctions, the lateral and basolateral compartments of epithelial cells are incompletely reached by the culture media. This is inhibitory to differentiation. For this reason, filter cartridges for the culture of epithelial cells were developed in the 1970s. These are hollow cylinders covered with a filter on one side (Fig. 3.6B). The filter cartridge is inserted into a culture dish. The cells are pipetted into the lumen of the hollow cylinder and are then able to grow on a filter, which simulates the conditions of a basement membrane. Completely different media can now be used apically and basolaterally for further cultivation. Since the apical and basolateral compartments exhibit small volumes, unfortunately liquid exchange between the upper and lower compartments takes place very quickly. In any case, no continuous gradient can be maintained in this way for longer periods.

Sample instructions for the culture of MDCK cells are given below. The following media are needed:

Freezing medium:

FCS	80%
Dimethylsulfoxide (DMSO)	20%

Culture medium for master cultures in plastic containers:

EMEM with 0.85 g/l bicarbonate	93 %
FCS	5 %
L-glutamine 200 mM in PBS	1 %
Penicillin/streptomycin	1 %

Medium for epithelial culture on filters:

EMEM with 0.85 g/l bicarbonate	88 %
FCS	10 %
L-glutamine 200 mM in PBS	1 %
Penicillin/streptomycin	1 %

When MDCK cells are in culture, they proliferate constantly, like tumor cells, and must be replaced in new culture containers after complete covering of the culture dish surface. If this is omitted, they die. This subcultivation of the master culture involves two steps.

(1) Preparation
Trypsin 0.05 %/EDTA 0.02 % in PBS without Ca/Mg, 10 ml per 750-ml culture flask, is preheated to 37 °C. Equipment needed: a 10-ml syringe with sterile filter attachment, a 50-ml beaker as a stand, a beaker for medium waste, PBS without Ca/Mg (sterile), culture medium (sterile and preheated), 75-cm^2 culture flasks, 10-ml pipettes (sterile).

(2) Execution
Completely aspirate old medium from the culture flask. Rinse two times with 10 ml PBS each; in other words, allow PBS from the pipette to run over the cells and remove it again. Add 5 ml trypsin/EDTA/PBS into the flask from the syringe with the sterile filter attachment. Loosen the cap of the flask and incubate for 15 min at room temperature, pour off, again add 1 ml trypsin/EDTA/PBS and incubate for another 15 min at 37 °C. The cells should become detached. After that, cells are observed under the microscope. Cells which potentially still adhere are brought into solution through light jarring of the flask. Add 9 ml culture medium and then determine the density of the cells in a counting chamber. The cell suspension is diluted in such a way that the resulting cell density is 1×10^4 cells/cm^2. This corresponds to 1×10^6 cells/ml in a 75-cm^2 culture flask. About 20 ml medium is pipetted into the new culture container. An aliquot of 1 ml of cell suspension is then added. The remaining cells are frozen or thrown out. After 3–4 days, the dish is once more completely covered with cells, which must then be subcultivated.

[Search criteria: continuous cell lines MDCK CHO]

3.2.4.3 Epithelial Cells in Functional Transfilter Experiments

Depending on the area of application, completely different cell lines are used in transfilter experiments. Apart from MDCK cells (Fig. 3.6), CaCo-2 and Calu-3 are used. The CaCo-2 line is a human colon carcinoma line that has been best shown to be used as an experimental model for intestinal absorption and has be accepted by the FDA as a

Fig. 3.7: Microscopy of cultivated tubulus cells of the kidney, which slowly form a confluent monolayer. Large gaps still exist between the individual cells. It is understandable that no functional transepithelial barrier is developed at this stage.

pharmacological/pharmaceutical *in vitro* model. The Calu-3 cell line descends from human bronchotracheal gland cells of the submucosa and is used as a model for the bronchial epithelia. The uptake of developed medications, proteins and DNA constructs can be examined on the intestinal or pulmonary cells. The cell lines can be acquired from the ATCC.

If the cells are allowed to grow on the membrane of a filter cartridge, then it is possible to investigate their growth, differentiation and development, as well as the maintenance, of an epithelial barrier. After the cells settle onto the membrane, they should build a functional barrier as fast as possible in order to investigate transport (Fig. 3.7). Although subculturing individual lines is done in a normal proliferation medium, special serum-containing culture media must be used for transfilter experiments, in most cases. Depending on the cell line, the medium contains 10–20% native or heat-inactivated calf serum, so that a confluent monolayer with intact tight junctions develops in the course of the culture.

Whether a functional barrier has developed must be determined electrophysiologically. In addition, the transepithelial electrical resistance (TEER) is typically measured with one measuring instrument and two electrodes, in the apical and basolateral culture media. The data show that it takes around 7–10 days until a TEER of over 1000 ohms/cm^2 is formed. However, a value over 500 ohms/cm^2 is indicative of an intact barrier. It should also be noticed that good TEER values are not achieved automatically, but depend largely on the respective culture conditions. This includes not only the filter cartridges, but also the sera used in the culture media.

Apart from the electrophysiological measurement of the TEER values, the sealing of an epithelium is frequently determined by radioactively labeled mannitol. It is important to note that mannitol is not taken up by the cells, but only over the paracellular pathway, between the two lateral plasma membranes, from the apical into the basolateral compartment.

To determine the tightness of an epithelium, a certain quantity of radioactively marked mannitol is pipetted into the apical culture medium. The amount of radioactivity present in the basal culture medium is measured after 1 h. If the epithelium is leaky, more radioactively marked mannitol travels paracellularly into the culture medium on the basal side of the filter. If less than 1% of the assigned radioactivity is measurable, it is an indication of an intact and thus optimal functionally developed barrier. If more radioactive mannitol is detected, it must be investigated whether it is due to insufficient development of the tight junctions, insufficient confluence of the cells or whether edge damage has occurred, in which the cells on the edges of the filter do not seal.

[Search criteria: epithelial cells transfilter culture]

3.2.4.4 Cultivation of Cardiomyocytes

A primary culture consists of cells isolated from an organ or a tissue and taken immediately into culture. In the production of cell cultures, the individual cells must first to be extracted from the organ or tissue. This is done by mechanical and enzymatic treatment of the tissues, as well as by means of special culture and growth conditions. In order to extract the cells from their tissue, the ECM is broken down with enzymes such as collagenase, trypsin, dispase or hyaluronidase. Afterwards, certain cells can be concentrated by gradient centrifugation or filter techniques. In addition, the cells can be mechanically separated from each other by the weak shearing stresses of careful vibration or pipetting.

   Based on a sample set of instructions, production of a primary culture from chicken embryos is now demonstrated. It consists of a preparatory step and a fairly extensive execution.

(1) Materials
   Preincubated eggs (approximately 8–10 days old), two sterile large curved forceps, one sterile medium-sized pair of forceps, one sterile small pair of forceps, one sterile medium-sized pair of scissors, one sterile small pair of scissors, sterile scalpels, sterile Pasteur pipettes, sterile Petri dishes (60 and 35 mm in diameter), sterile metal egg cups, more sterilized 100-ml Erlenmeyer flasks with ground glass stoppers, sterile small magnetic stir bar, sterile centrifuge glasses, cell counting chamber, PBS, 0.25% trypsin in PBS, FCS, MEM, Trypan blue.

(2) Execution
   The incubation of the fertilized eggs takes place at 385 °C and a relative air humidity of 60–70% in a special incubator. The preincubated eggs are taken under the sterile hood and opened. In addition, the eggs are placed into a holder with the larger end upward and cleaned carefully with 70% ethanol.
   With curved sterile forceps, the egg is broken open and a round opening is made in the shell. Afterwards, the outside white membrane is removed. The embryo is now visible and is lifted out with the large curved forceps. The embryo is transferred into a 60-mm Petri dish containing ice-cold PBS. The head is cut off with large shears and the chest area is opened with small shears or a scalpel. The beating heart is removed with small forceps and put into a Petri dish with ice-cold PBS. Between 10 and 15 (maximum 20) embryos are prepared in such a manner.

When all the hearts are removed, the large blood vessels with stumps are removed from each heart. Afterwards, the hearts are washed twice with ice-cold PBS solution. The hearts are then put into a volume of 1–1.5 ml PBS and cut into as small as possible pieces with two scalpels. The pieces are transferred with sterile Pasteur pipettes into an Erlenmeyer flask and suspended with 5 ml 0.25% trypsin solution in PBS. The cells are then incubated for 10 min at 37 °C under gentle agitation. The trypsin supernatant is then removed with a sterile Pasteur pipette and thrown out. The bits of heart remaining in the Erlenmeyer flask are incubated again with 5 ml trypsin solution at 37 °C for 10 min and under weak agitation. Afterwards, the supernatant is removed with a sterile Pasteur pipette and transferred into 2 ml of FCS in order to block protease activity. Following 5 min centrifugation at 1300 r.p.m., the supernatant is thrown out. The sediment is taken up into growth medium (85% MEM/15% FCS). The pellet is swirled up into the growth medium and the resulting cell suspension is then placed on ice. The trypsinization is repeated twice using the remaining, undigested pieces of heart. The resulting cell suspensions are now combined and mixed very well. Next, the cell number is determined. This should be around 0.5×10^6 cells/ml. With a higher cell density, the culture is diluted accordingly with medium. The cells are sown in Petri dishes or culture flasks. Culturing is done in the CO_2 incubator. After 1 day, the first medium change is made. The feeding medium is 90% MEM and 10% FCS. Heart muscle cells begin to adhere to the culture dish after a few hours and rhythmically contract after 48 h. With the appropriate working standards, it is also possible to work without antibiotics.

With all of these dissociation or disintegration experiments, in which the help of a protease is used for cell isolation, it is important not forget that not only the pericellular matrix, but also the cells themselves are open to attack. Thus, trypsinizing for too long, for example, can exert a toxic or even lethal influence on the isolated cells. This shows itself in a bad vital yield and poor growth.

Apart from the enzymatic treatment, one frequently works with buffer systems that are poor in, or free of, calcium and magnesium. The lack of calcium and magnesium leads to a softening of the cell adherence, and finally to a separation of the cells. Frequently, the chelating agent EDTA is added to these media in order to make the Ca^{2+} unavailable. The dissociation times with this method are significantly longer than with than with enzymatic processes. In support, the tissue should be carefully taken up several times through thin Pasteur pipettes. In this way, excellent cell suspensions can be obtained by use of easy shearing stresses.

The process presented makes it possible to isolate different cells from a tissue without too much destruction to the integrity of the cells. This protocol is not suitable for all tissue and optimal conditions need to be determined experimentally for each tissue. The results are considered satisfactory if a high yield of growing cells is achieved and the viability is more than 90%.

Since organs and tissues consist of varied cells, the question of cell purity arises after the isolation of cells. It must be clarified whether only one cell type is to be taken into culture with the preparation or whether the culture is to consist of several cell types.

How these different cells can be separated is not discussed here in detail; nevertheless, this process must be considered individually and very critically. The complication in working with a mixture of many different cell types is that not all cells grow at an equal rate in culture. For this reason the overgrowth of an individual cell type can occur very easily over time. Turned around, this phenomenon can naturally be used to quite simply allow a fast proliferating cell type grow within a short time, in large quantity and in pure form. If, however, one is interested in slowly growing cells, only special techniques such as using a cloning cylinder or special selection media can help in isolating this cell type.

Finally, it should still be noted that adult cardiomyocytes develop from cardiomyoblasts which, just like neural cells, belong to the postmitotic cells. For this reason, these cells cannot be stored or subcultivated and no cell lines can be made from them. Therefore, in pharmacological investigations, cells must be isolated and taken each time from the organ into culture, depending on the requirements of the experiment.

[Search criteria: primary cell culture cardiomyocytes isolation]

3.2.4.5 Cryopreservation

Cells are cultivated only when they are needed, since their maintenance costs a great deal of time and money. Many cell types can be frozen and thawed again if necessary without any problem. During freezing, the cells are protected from the formation of intracellular ice crystals by a solvent. The cells can be stored frozen for nearly arbitrary periods. In each case, the cells should be present as a suspension. In addition, cells such as MDCK cells are dissociated with trypsin and resuspended in normal growth medium at a concentration of $2-4 \times 10^6$ cells/ml. The cell suspension is then cooled in an ice-water bath. Immediately afterwards, sterile glycerol or DMSO at a final concentration of 10% is added. With a wide diameter needle and a sterile syringe, 1 ml of the cell suspension is transferred into a sterile glass ampoule, which is closed immediately. Alternatively, there are also numerous freezing tubes (cryotubes) made of plastic with screw caps. The closed tubes are then placed into a polystyrene box. Such boxes are frequently used for flask transport. The box should have a wall thickness of 5–10 cm. With the fitted cover, the box is closed and placed into a – 70 °C freezer. One can estimate that the material cools by around 1 °C/min. After approximately 2 h, the tubes are transferred into liquid nitrogen, arranged in the storage container and labeled in a clearly organized protocol. The cells can be kept for many years.

The cells can be revived again from their cooled sleep as required. To thaw out frozen cells, the cryotube is taken out of the nitrogen container. The cryotube is then transferred very quickly into a 37 °C water bath, in which the sample is thawed and kept at a moderate temperature. Afterwards, the tubes are wiped off well with 70% alcohol and opened. The sample is transferred into an appropriate culture container under a sterile hood using a sterile pipette. Growth medium must now be added in order to dilute the glycerol or DMSO contained in the sample. If no cryoprotection at all is to be contained in the sample, the cells must be centrifuged at 800 r.p.m. for 5 min. The supernatant is removed and replaced by new pre-warmed growth medium.

Tab. 3.5: Problems that can occur in culturing cells.

Difficulty	Source of problem	Solution
Cells do not adhere to the surface	possibly left in trypsin too long during cell detachment	shorten the exposure time to trypsin, as well as reducing the concentration
	infection with mycoplasma	test for mycoplasma infection
	appropriate binding factors are missing from the surface	coat the surface with peptides or ECM proteins
Visible precipitate in the medium, without a change in pH	detergent residues can cause precipitates; thawing of frozen medium	clean effectively with less detergent and multiple rinsings with deionized or distilled water
Precipitate with a change in pH	bacterial or fungal infection	attempt to decontaminate cultures with antibiotics and antimycotics
Change in pH of medium	cells are too dense	decrease cell density; correct CO_2 level
	incorrect CO_2 partial pressure, too little bicarbonate buffering	adjust bicarbonate concentration, addition of addition buffer
	closed culture flask	improve inhibited ventilation
Reduced growth of culture	medium and additives are from different producers	critically compare the contents of old and new batches
	insufficient or degraded essential components	check glutamine content
	few cells visible	increase the number of cells
	possible contamination with bacteria or fungus	investigate medium through a microbiological lab
	check the number of passages	aging culture
	missing CO_2 ventilation	cell death
	check the gas bottle and regulation of the incubator	presence of toxic metabolites, antibiotics or other additives

The cells are then kept in the CO_2 incubator at 37 °C for 24 h. Afterwards, the medium should be changed.

[Search criteria: cell cryoconservation freezing media]

3.2.4.6 Problems with the Culture

Many problems can occur with the culture of cells (Tab. 3.5). Completely different difficulties are encountered with continuous cell lines than with primary cells.

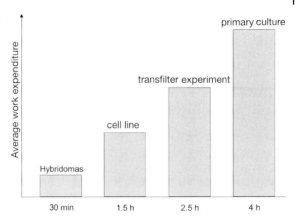

Fig. 3.8: Comparison of the work expended with different cultures. Hybridomas, cell lines and primary cultures need completely different material and time expenditures in terms of isolation and subcultivation.

3.2.4.7 Work Expended with Cell Culture Work

The previous examples show that there are not one, but various kinds of cell culture (Figs 3.2 and 3.5). With the simplest method, one multiplies non-adherent cells in a culture medium alone by pipetting (Fig. 3.2). Work with adherent cells of cell lines is more complex since these need to be detached from the growth surface, isolated and counted before the pipetting steps (Fig. 3.5). Primary cultures are the most work intensive, e.g. the isolation of cardiomyocytes, which are isolated in several steps from an animal or an organ and then brought into culture.

Experimental experience shows that the daily maintenance of a hybridoma line requires only about 30 min, while adherent cell lines take around 1.5 h and production of primary cultures takes approximately 4 h (Fig. 3.8). The different work expended in the production of individual cell cultures has already been described. It is important to

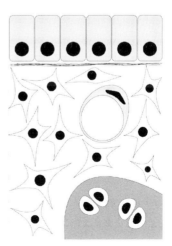

Fig. 3.9: Schematic section of a tissue explant. On top is an epithelium, which is separated from the connective tissue under it by a basement membrane. Between the loosely distributed polymorphic connective tissue cells is a capillary for the necessary nutritional supply. On the bottom is hyaline cartilage.

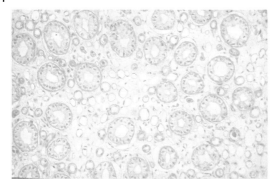

Fig. 3.10: Histological cut through the medulla of the kidney as example of a section of tissue. Note the collecting ducts with large lumens and other tubule or vascular structures, which are surrounded by interstitial connective tissue.

consistently follow a culture protocol. For this reason, it is not possible to accelerate the process and still produce high-quality cultures.

3.3
Tissue Culture

Cultures with animal and human cells have become indispensable tools in biomedical research and therapy. Cell cultures which divide quickly and produce monoclonal antibodies or form recombinant proteins are in most demand. With this technology, cells grow as isolated as possible from their neighbors on the surface of a culture container. They show few, if any, characteristics of tissue structures.

Histological preparations show that pieces of organs or tissues, even if they are small, are very complex in their composition and usually consist of several rather one individual, homogenous, tissue. Apart from the specialized cells of the respective parenchyma, blood vessels, fibroblasts and the immunological defense are found in the stroma (Figs 3.9 and 3.10).

Culture experiments with tissue begin with about 500-μm thin pieces of tissue from brain, liver, kidney, pancreas or an artery, and are sterilely removed and inserted into a culture dish. In order to supply the respective explants, culture medium is added, usually containing FCS. The entire culture dish is not filled with medium, but just enough is given so that the explant is covered with medium. This is to prevent the explant from floating. Using this method, oxygen can diffused over a short distance to the cultivated tissue.

Tissue culture must be seen differently than cell culture in that tissues consist of socially organized cellular networks. Instead of cells proliferating as fast as possible, tissue culture explants are held in an original condition for as long a time as possible under *in vitro* conditions. This seems very simple; however, for many reasons, the optimal tissue culture resulting in perfect function has not yet been successfully accomplished. For decades it has been discussed whether tissue culture really continues life or only prolongs cell death.

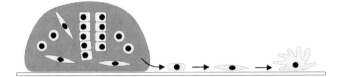

Fig. 3.11: Example of the cell of cells in a tissue explant onto the surface of a culture dish after several days of culture in a serum-containing medium. During the culture, numerous cells from the explant migrate to and grow on the culture dish surface. A complete reorganization takes place within the explant.

Frequently, organ culture is included in the area of the tissue culture. A goal of organ culture, however, is not the maintenance of a tissue structure, but to lead available embryonic tissue, under *in vitro* conditions, into normal development in order to observe the physiological development in the maturing organism. Today's tissue engineering takes an intermediary position, since it contains aspects of both cell and tissue culture as well as organ culture.

[Search criteria: tissue culture organ culture]

3.3.1
Migration and New Formation

If a heterogeneously built-up tissue explant is exposed to a serum-containing culture medium on the surface of a culture dish, it is in principle to be assumed that all cells will survive in the first days. The explant, however, changes its outer and internal structure within hours. For reasons not yet clarified, a high percentage of cells begin to abandon the explant and move into the periphery or onto the surface of the culture dish, either immediately or after days. Sometimes the entire explant will rearrange. First macrophages and leukocytes appear, followed by fibroblasts. Finally, migrating epithelial cells become visible (Fig. 3.11).

There are two possibilities for the migrating cells of a tissue explant culture. If the explant is kept floating in the culture medium, the cells on the surface will move along

Fig. 3.12: Microscopic representation of a tissue explant on the culture dish surface. The majority of cells does not remain in the explant, but migrate and form a monolayer.

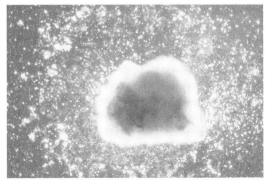

and remain in direct contact with the tissue. This migration shows that only one part of the cells migrates, while the other part remains inside the explant. The explant is then covered with an epithelium or, in most cases, with fibroblasts. However, if the explant has contact with the bottom of a culture dish, then a large part of the cells will wander along the surface of the culture dish. These cells can be further bred as a monolayer after removing the tissue explant.

The migration of cells out of an explant can be used experimentally. Frequently, tissue explants are taken into culture in order to yield individual cell types from the smallest samples. The advantage of this method is that the cells can be obtained by migration and thus without breaking down the tissue with proteases (Fig. 3.12). A disadvantage of this method is that most of the migrating cells lose functional characteristics due to their dedifferentiation.

[Search criteria: cell migration tissue organ culture explant]

3.3.2
Dedifferentiation

Over the years, many different tissues have been cultivated using serum-containing media in culture dishes. In all of our culture experiments, cells both remaining within the explant as well as those that had migrated out showed strong changes within only a few days. The migrated cells left areas that were restructured and revived. However, the areas that developed again only resembled the original form and function in a few cases. For this reason, sections of tissue such as brain, stomach, liver, pancreas, kidney, blood vessels or diverse connective tissues cannot be held in culture for longer periods without the loss of many of their typical characteristics. The changes of morphologic, biochemical and physiological characteristics during culture are called cellular dedifferentiation. Up to now there is no known cultivated tissue that does not exhibit changes through cellular dedifferentiation during long-term culture.

A typical example of cellular differentiation is the culture of renal glomeruli. These can be isolated from the kidney by filter techniques and gradient centrifugation (Fig. 3.13). Each glomerulus has a diameter of $100-130$ μm dependent on the species

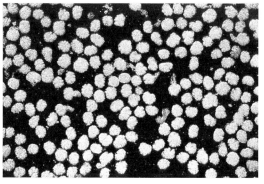

Fig. 3.13: Microscopic representation of isolated glomeruli of the kidney on the surface of a culture dish.

Fig. 3.14: Microscopic representation of a renal glomerulus after several days of culture in serum-containing medium on the surface of a culture dish. When the cells are grown, it is not recognizable whether they are podocytes, mesangium cells or endothelial cells.

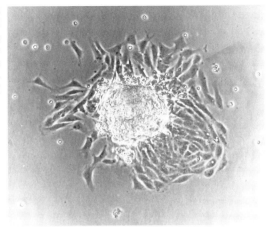

Fig. 3.15: Example of the cellular dedifferentiation of cells on the surface of a culture dish. On the basis of the white granula, it can be clearly shown immunohistochemically that not all cells still form a certain protein. In the tissue, however, this protein is not found in the individual granula (picture center), but in the entire cytoplasm.

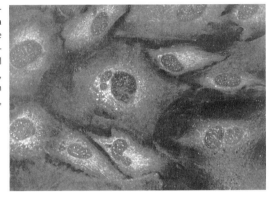

and consists of several cell types, including podocytes, intraglomerular and extraglomerular mesangium cells, and endothelial cells.

If glomerular cells are isolated and cultivated in serum-containing culture media, after some days it can be observed that the cells have covered the dish and have spread as a monolayer into the periphery of the culture dish (Fig. 3.14). It is no longer recognizable whether the cells are podocytes, mesangium or endothelial cells. In intact glomeruli of the kidney, in contrast, these cells look very different from one another and are easily differentiable by eye.

A further example can be seen with epithelial cells from the collecting duct of the kidney, which migrate from an explant culture and grow on the surface of a culture dish (Fig. 3.12). The cultivated cells grow untypically flat and polygonally. In addition, they grow a remarkable distance from neighboring cells. Immunohistochemical evidence has shown that only few of these cells in the center of the image are able to still form a protein typical of the tissue (Fig. 3.15). The synthesized protein is visible in the form of white granula. In contrast, most other cells in the periphery do not show an

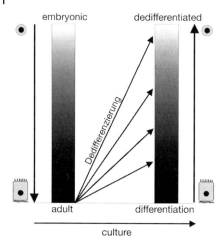

Fig. 3.16: Schematic representation of development and dedifferentiation in tissue culture. During development, functional tissues develop from embryonic cells. This process is called differentiation. If functional tissues are brought into culture, then morphologic, physiological and biochemical characteristics are lost to different extents. This process is called dedifferentiation.

immunohistochemical reaction and have lost the ability to produce collecting tubule-typical proteins during culture due to cellular dedifferentiation.

The extent of problems with cellular dedifferentiation is best explained on the basis of cell patterns (Fig. 3.16). During embryonic development, the cells in an organism mature to functional tissue cells with very specific characteristics. This process, naturally carried out, is called differentiation. If functional tissues are taken into culture, typical characteristics are only partially retained or can even be completely lost. Depending on the type of tissue, not all characteristics are lost equally. Neither the migrated cells nor those remaining in an explant maintain their original morphologic and functional condition.

In try to avoid dedifferentiation, the conditions must be selected for the tissue culture in such a way as to prevent development into non-specialized cells and maintain the functional characteristics as much as possible. Also of note is that tissue culture cells exist in different developmental stages, which can be neither functionally adult nor purely embryonic.

The reasons for cellular dedifferentiation under culture conditions are varied. For example, an isolated tissue explant is missing a functional vascular system. It does not possess a satisfactory disposal system for metabolites and experiences no neural control. A tissue in the adult, functional condition, on the other hand, is organized in such a way that cells are present there after reaching a certain size, in a certain density. Depending on the organ or tissue, there are also cells to be found that divide very frequently, while other directly neighboring cells hardly divide at all, despite the same environment. These natural control mechanisms are removed in tissue culture. Practically all cells existing in a tissue explant are reprogrammed under culture conditions by the addition of FCS and stimulated for migration, as well as cell division. Because of this, they leave their traditional environment and partially rearrange the explant or can be cultivated as a monolayer on the surface of a culture dish.

The causes of cell migration out of a tissue explant are also varied. One of the main causes is the medium used. Most culture media were developed 40–50 years ago for a very specific problem. At that time, the intention was not to cultivate a tissue, but rather individual cells in the form of a monolayer. These cultures are meant to grow as fast as possible, in order to most efficiently produce viruses. The origin, appearance and further characteristics of these cells were completely unimportant. Accordingly, the culture media were optimized in terms of their electrolyte content and nourishing factors in such a way that they supported the fast proliferation of cells. This is the reason that cells are torn from their natural balance between interphase and mitosis by the application of such a culture media and begin to permanently divide.

In addition, FCS is frequently used automatically when setting-up culture media. Mainly it is used to support the fast proliferation of cells. This is due to the content of mitogenic factors, which lead the cells as rapidly as possible from one mitosis cycle to the next. In the intact tissue network, however, the cells are subject to individual controls of division that are apparently inactivated by the isolation of the explant.

In addition, FCS and spreading factors are present which stimulate the cells to move and distribute themselves. The coordination of the mitotic stimulation and the spreading activity causes tissue cells to lose their consistency of location in a cultivated explant, enabling growth. One can observe the migration of cells in many slice cultures in which thin sections of tissues or whole organs such as the neural hippocampus are taken into culture, in order to perform physiological or pharmacological/toxicological experiments with them. The cells remain bound to their location and maintain their typical function for a relatively short time of some hours. Then, irrevocably, the migration of cells begins along with the reorganization of the tissue and simultaneous dedifferentiation of the cells.

[Search criteria: dedifferentiation culture loss differentiation]

3.4
Organ Culture

By definition, tissue culture and cell cultures are differentiated from one another. Organ cultures come from removed organ anlages, regenerating adult organs or parts of them. Organs consist of multiple tissues. The fact that cell differentiation and the histo-architecture, as well as the overall function of the respective organ with its individual tissues, is retained as much as possible and potentially further developed during the culture phase is dependent on the organ culture. The use of organs of embryonic, fetal or perinatal origin, such as lung, liver, salivary gland or kidneys, whose further development one would like to observe in culture, is preferential for this kind of the culture.

The first valuable information on organ culture was gained through experiments with explants. Here, it was shown that by combining tissues, such as the spine and kidney mesenchyme, embryonic cells introduced could be stimulated into tissue development or maturation. On the other hand, these findings also showed that mature, functional tissues did not automatically develop in this way under *in vitro* conditions. Through exact analysis, constructs which exhibit a broad spectrum of embryo-

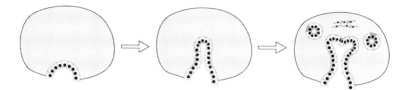

Fig. 3.17: Illustration of an embryonic organ that develops by branching morphogenesis. An epithelium bud grows into the mesenchyme. The epithelium bud divides several times, forming a branched tubular system, from which the later parenchyma develops.

nic to adult characteristics result. Particularly important in this context is the fact that not only proteins typical to the tissue, but also atypical and thus foreign proteins can be expressed.

Embryonic tissue behaves differently in organ culture than adult tissue structures, which can be easily explained based on the experimental examples of branching morphogenesis (Fig. 3.17). This process is found with the development of glandular organs, which are composed of a parenchyma and a compartmentalizing stroma. The typical development of these organs consists in an epithelium bud growing into an embryonic connective tissue (mesenchyme). The epithelial tube extends itself, a lumen is formed and finally regularly returning branches are built. Thus, a branched duct system develops. The gland epithelia end pieces, lying in the connective tissue, develop secondarily to the actual functional epithelium surrounding them. According to this pattern, organs such as the liver, pancreas, salivary glands and the kidney develop.

The developmental physiological aspects, which, on one hand, lead to the master formation of the duct system and, on the other hand, to the functional development of the glandular end pieces, with their special secretory function, are of great interest. It has been possible for decades to accomplish this experimentally with embryonic organ cultures. In addition, the organ anlages are removed both intact and sterile, and brought into culture. Since this embryonic tissue exhibits cells with a high capacity for proliferation, the experiments can also be accomplished very well with serum or growth factor-containing culture medium. The organ can, however, only be removed up to a certain size. The limitation is that due to the lack of blood circulation, an insufficient supply of oxygen and nutrients takes place at a certain tissue thickness. Consequently, partial death of the tissue occurs on the inside of the culture after some time.

[Search criteria: organogenesis branching morphogenesis organ culture]

4
Tissue Engineering

The current field of tissue engineering was essentially founded by Charles and Josef Vacanti as well as Charles Patrick, Antonios Mikos, Robert Langer and Larry McIntire in the 1980s. There now is almost no branch in biomedicine that does not deal with this new discipline. The goal of tissue engineering mainly consists of activating regenerative abilities of the body that have come to a standstill and, if necessary, replacing damaged tissues with tissue implants (Fig. 4.1). Due the special technical difficulties, the generation of tissue constructs requires the particularly close cooperation of medical doctors, cell biologists, material scientists and engineers.

Therapy with cultivated cells (Fig. 4.2 and Tab. 4.1) is distinguished from the production of tissue constructs (Fig. 4.3 and Tab. 4.2) and the building of organ modules (Fig. 4.4 and Tab. 4.3).

The spectrum of tissue engineering covers all kinds of tissue present in the body. A host of technical methods for constructing tissue or even organoids from cultivated cells are also included in "technical engineering". Partly, it refers to functional cell or tissue constructs that are used as biological implants for patients and partly as technical biomodules at the bedside.

Present-day transplantation of organs and tissues from a donor to a recipient always leads to chronic rejection reactions. For this reason, the rejection reaction must be

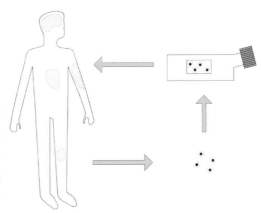

Fig. 4.1: Principles of cell therapy and tissue engineering. Cells from the patient are isolated and brought into culture for the regeneration of damaged tissue. These cells, or tissue constructs, are implanted into the damaged area, such as the brain, heart, bone or muscle, to trigger healing processes.

are injected with as little volume of liquid as possible. The hollow is closed with a piece of periosteum, and then the edges are sewn and sealed by fibrin glue (Fig. 4.2). Cultivated chondrocytes can also be used for sealing. They are mixed with fibrin or agarose and applied across the damaged area. Protected in this hollow, the cells should now mature to hyaline cartilage.

By definition, implantation of cultivated chondrocytes is not considered tissue engineering, but as cell therapy, since isolated cells are used for implantation in this case and not a tissue construct. The chondrocytes in this form are not settled onto a matrix, have not produced cartilage ground substance and are injected as a cell suspension in media of different viscosities with a syringe into a prepared hollow of the periosteum in the damaged joint surface. It is only within this hollow that real tissue production takes place and thus, over time, generation of mechanical stress-resistant cartilage ground substance.

In another strategy, autologous chondrocytes are cultivated within a scaffold for a few days. The cells have the chance to adapt to the scaffold, settle and produce ECM proteins. The construct, which is still flexible, can be fixed to the damaged joint surface and heal without needing to be covered by a piece of periosteum. By definition, this method is not cell therapy, but rather the implantation of a tissue construct (Fig. 4.3). The probability of losing cells is relatively small. Finally, it is possible to generate a cartilage construct for weeks under *in vitro* conditions until it shows certain mechanical resistance. Before the implantation, the construct is cut to the right size and placed within the damaged joint surface into which it should heal as soon as possible and contribute to the mechanical stability of the joint surface. However, the integration of the construct into the surrounding tissue is still relatively problematic.

[Search criteria: autologous chondrocyte transplantation]

4.1.3
Large-scale Burns

Apart from bone marrow transplantation or leukemia, the longest clinical experience in the area of cell therapy exists mainly in the therapy of patients with the most serious burns, whose lives have been saved by cultivated keratinocytes. There are many instances of accident victims with burns on more than 90 % of their body surface. Apart from basic care for these seriously wounded patients, keratinocytes from undamaged areas of the axilla, groin or the foreskin are isolated and placed in culture. The keratinocytes are then multiplied within special culture flasks with a removable cover, mostly on a layer of fibroblasts (feeder layer). It is only recently that cells are being grown on a synthetic ECM, which provides mechanical stabilization during transplantation. You can imagine how many culture flasks with a base area of approximately 120 cm^2 are needed in the form of many patches to cover the burn surface area of a patient. Another application is the simultaneous placement of cultivated keratinocytes with fibrin glue onto the burned skin surface. In addition, great success has been achieved in the therapy of large (crural) ulcers and decubital ulcers.

By definition, the therapy of large burns described is not transplanted tissue, but proliferated cell cultures of keratinocytes. Single keratinocytes are isolated from the remains of skin and amplified until a sufficient cell mass has been reached, so that the area of growth is sufficient to cover the wound. It should be taken into consideration that functionally important secondary formations of the skin, like reserve folds, hair, sweat and sebaceous glands, cannot be regenerated with the methods used today, which leads to distinct disabilities and limitations in the patient's quality of life.

[Search criteria: burn skin keratinocytes tissue engineering]

4.1.4
Muscular Dystrophies

Muscular dystrophies are marked by a progressive loss of skeletal muscle and belong to the group of frequently deadly hereditary diseases. Despite great advances in the identification of mutated genes, the possibilities of therapy are very limited. Potential future opportunities are offered by cell therapy and tissue engineering.

The origin of muscular dystrophy is traced back to the *dmd* gene, which is located on the short arm of the X-chromosome. With its 79 exons and 2.5 megabases, it is one of the largest known genes. It codes for the cytoskeletal protein, dystrophin, with a molecular weight of 427,000 Da, which is located on the inner layer of the cellular membrane. Its N-terminal end is connected to actin filaments and its C-terminal end to a dystrophin-associated glycoprotein complex (DGC). This intracellular complex itself is connected through the plasma membrane to the basal lamina, by the matrix proteins laminin and agrin. In this way a mechanically strong connection between the cytoskeleton of the muscle fiber and the basal lamina that surrounds the fiber is created in skeletal muscle. The genetic changes in different components of the DGC lead to different types of muscular dystrophies. Mutations in the laminin-2 gene are the molecular cause for congenital muscular dystrophies, while changes in the different sarcoglycans cause limb–girdle dystrophies.

The changes within the dystrophin complex result in an interruption between the cytoskeleton and the basal lamina. Thus, the cellular membrane rips open during contractions. This again leads to a degeneration of the muscle fiber. The course of Duchenne's muscular dystrophy shows possible regeneration during the younger years, which decreases with age. This leads to a loss of muscular mass, followed by the loss of whole muscle groups.

Neither transfer of myoblasts nor gene therapy have been successful in achieving improvement, let alone a cure, in clinical studies. Serious problems during therapy are posed by the body's own immune system, which is directed against the dystrophin molecule. For this reason attempts are being made to make use of compensatory mechanisms that do not result in an immune reaction. Therefore, the expression of proteins already available should be amplified. This is attempted by overexpression of utrophin, for example. This is a molecule that is structurally and functionally closely related to dystrophin. Also, overexpression of agrin could replace the mutated laminin-

4.2
Tissue Constructs

There are great advantages for regeneration and healing if whole functional tissues or their mature precursors, instead of isolated cells, can be use with patients (Fig. 4.3 and Tab. 4.2). To do this, cells are isolated from the patient, e.g. from unburdened cartilage, and multiplied in culture to reach a sufficient cell mass. Alternatively, stem cells can be used here.

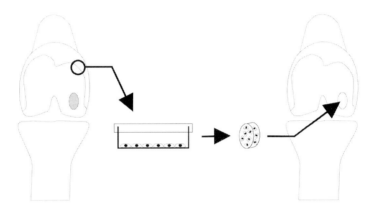

Fig. 4.3: Principle of artificial tissue creation. The creation of a cartilage construct with cultivated cells and an ECM (scaffold) for therapy of joint damage is shown.

Tab. 4.2: Uses of tissue constructs for the regeneration of lost function in the basic tissue types.

Tissue replacement	Disease	Use	Problems
Neural constructs	separation of the spinal cord, degeneration of the retina	implantation	insufficient differentiation
Muscle constructs	lack of muscle tissue	implantation	insufficient function, inadequate connection
Epithelial constructs	esophageal cancer, urinary bladder cancer, corneal damage	implantation	separation, not tight, insufficient function
Connective tissue constructs	damage in form giving cartilage, comminuted fracture of bones, measures of reconstruction	implantation	mechanical stability, stability of form
Blood vessels	aneurysms, arteriosclerosis	implantation	mechanical stability
Heart valves	defects of the heart valves	implantation	calcification, lacking function

When the desired cell mass is reached, the cells are transferred onto an artificial ECM. Within or on top of this scaffold, the cells already start to develop a functional tissue in culture. Consequently, at the time of implantation, a partly mature tissue is already available, through which the risk of an undesired development is reduced and the duration of the healing process is shortened. Most of the time, the handling and the mechanical resistance of such a construct are better than with cell therapy.

The following are some examples of the clinical applications of tissue constructs (Tab. 4.2). Here, distinct limitations are set by insufficient knowledge in the area of tissue development. Often the constructs do not reach the intended differentiation or level of function needed.

[Search criteria: tissue engineering artificial constructs]

4.2.1
Defects in Structural Connective Tissue

Mechanically stressable constructs of cartilage and bone, which could be used in joint damage or for treating osteoporosis, are in particular demand. Artificial cartilage tissue can be produced by applying chondrocytes onto a suitable matrix (scaffold) under culture conditions. The cells have to spread out evenly, adhere and generate the typical ground substance for cartilage during culture, which lasts for weeks. The matured tissue construct can be removed with forceps, cut to the right size and shape, and implanted (Fig. 4.3). Similarly, isolated osteoblasts can be use for bone production, and fibroblasts for the production of tendon and ligament constructs.

[Search criteria: cartilage scaffold tissue engineering]

4.2.2
Bones and Fractures

Constructs of hyaline cartilage do not need to be thick, since this tissue naturally covers the joint surface with a layer only 1–3 mm thick, is nurtured by diffusion and consequently does not require its own functional support by blood vessels. In contrast, artificial bone constructs are required with a thickness of centimeters for therapy. However, such massive constructs require their own vascular support, which is not yet available under culture conditions. Additional difficulties occur after the implantation of huge tissue constructs. Most of the time, the vascular system cannot grow into it fast enough. Thus, the implant is undersupplied and the cells die. It is for this reason that only relatively small and thin tissue constructs of bone fibers can be created.

[Search criteria: bone repair tissue engineering]

4.2.3
Reconstructive Measures

Next to support tissue, the production of loose connective tissue is of great biomedical importance after surgical operations. Loose connective tissue and fat tissue are necessary for the reconstruction of areas from which tumors of the salivary glands have been removed, for example. At this point, problems are not only posed by the production of the actual tissue construct, but also by the required functional vascular connection. Implants with a thickness of many millimeters to centimeters are required for filling. Technically, it is still impossible to create such constructs under culture conditions, as the cells within the deeper layers cannot be supplied with nutrients and oxygen by diffusion. If such thickly layered constructs were implanted, the cells in the interior would die before their own capillary system could be developed. Such necrosis negatively influences regeneration and can easily be infiltrated by bacteria.

Storage and structural fatty tissues can be distinguished according to their physiological importance. During fasting, storage fat is dismantled over time as an energy-rich reserve. Structural fat remains mostly unchanged with this type of fat reduction. Structural fatty tissue is located as bodies of fat in the joints (corpus adiposum infrapatellare), within the orbita (corpus adiposum orbitae), in the cheek (corpus adiposum buccae), and as pads in the heels, the sole of foot, the palm and the buttocks. It often occurs in locations where tissue atrophies. This includes the thymus, bone marrow and muscles. Structural fat fills the space where the milk gland will development in the female mammary gland.

With regard to tumors of the large salivary glands and breast carcinoma, radical operations and irradiation often have to be performed to remove the necessary tissue. Large hollows remain that then have to be filled with fat and loose connective tissue. In order to fill connective tissue spaces after radical operations, thick-layered tissue constructs are necessary. For physiological reasons the constructs with a certain thickness cannot be nurtured by diffusion only, but require connection to a vascular system. From a technical and cell biological point of view, the generation of artificial fatty tissue corresponding to these requirements is still far away.

[Search criteria: breast fat tissue engineering]

4.2.4
Damage to the Cornea

Tissue engineering with epithelia is an unexpectedly wide field. All projects with epithelia have proven to be especially difficult during their experimental realization. The reason is that cultivated epithelial cells require a special surface for their fixation, which both guarantees the stability of the basal lamina and has a positive influence on the differentiation of the tissue. One may think that these problems should already be solved with the numerous biomaterials today available. However, this is not the case or is at least very restricted. It is possible to foresee the technical and cell biological difficulties presented by bringing epithelia into contact with new biomaterials. Reactions that could not be predicted frequently occur. Most of the time there are strong morphological, physiological and biochemical changes in the epithelial cells. Consequently, the cells lose their transport and barrier functions, can no longer withstand rheological stress, and disconnect from the surface.

As an example, experiments in ophthalmology for the regeneration of the cornea are presented. A cornea which has been damaged in an accident is always connected to severe impairment of the eye. Thus, the only possible therapy for patients with chemical burns, non-healing inflammations and scars is the transplantation of a donated or artificially created cornea. Such a keratoprothesis consists of an optically effective part (optic) and an anchoring part (haptic), which is fixed by the growth of fibroblasts and the production of an ECM.

Despite intensive research in this area, the cell biological problems of keratoprotheses have not been satisfactory solved. An essential requirement in the generation of a keratoplasty is that it corresponds to a transparent and optimally epithelialized stratified tissue. For this to happen, the tissue has to optimally grow over the whole ventral side of the implant. In addition, the cells next to the basal lamina need to have stem cell characteristics. Step by step, they provide a continuous regeneration of the stratified epithelium. They need to differentiate during the regeneration and must not lose their transparency in the process.

The artificial ECM, on which the keratinocytes should be placed, poses experimental problems during the creation of a keratoprothesis. Just as important is that the transparent biomaterial used should exhibit the optimal differentiation, stays transparent and produces a tissue-specific, or transparent, matrix. This means that the choice of available matrices is very small. The optimal covering of the biomaterials used and the profile of differentiation in the corneal epithelium generated in culture must be determined through extensive investigations.

[Search criteria: cornea tissue engineering]

4.2.5
Tumors of the Digestive System

The epithelia covering the lumen of the digestive system is permanently renewed. The cellular source of the renewal is the stem cells located in the upper area of the glandulae gastricae of the stomach, and in the lower area of the crypts of the small and large intestine, that provide a life-long supply of differentiated cells.

Inflammatory and necrotic processes, like tumor diseases, can lead to large-scale ruptures in the esophagus, wall of the stomach and intestine. For the patient, this means that relatively large areas of the particular organ have to be removed, since the rupture cannot be covered with suitable material. Problems are posed by the complex morphology and functionality of the natural organ wall.

A functional tissue construct for the wall of the esophagus, stomach and intestine must include the layers of the tunica mucous, the tela submucosa and the tunica musculuaris. This also means that the layer of epithelium which covers the lumen must be polar, include the necessary cell types, be able to regenerate and can withstand the natural mechanical stresses. In the tunica submucosa, the required blood vessels and nerves should to grow as fast as possible. Furthermore, the tunica muscularis needs to have enough smooth muscle for the necessary peristalsis.

Up to now, no one has succeeded in generating such complex organ parts like the wall of a functional digestive organ. For this reason, an alternative strategy is followed. Small intestine submucosa (SIS), among other things, is used for covering ruptures in the digestive system. These are preparations of the reticular matrix in the small intestine. The intestine wall is treated with different detergents and freed of cellular debris. A three-dimensional lattice network made of fibers is left behind and then sewn onto the rupture. The hope is that cells from all typical wall layers will migrate into the matrix and, over time, build a layered regenerate within the SIS, using the concept of guiding.

[Search criteria: SIS bowel tissue engineering]

4.2.6
Sick Blood Vessels

Large and small blood vessels can become ill during the course of their lives, so that their walls change and the blood supply is inhibited. A typical example can be seen with restricted coronary vessels. A bypass operation may then be necessary. Veins from the leg, which usually are not exposed to systolic blood pressure, are taken in order to bypass the restricted location. It would be ideal if tissue constructs from the body's own cells could be used, which would be available in any length, any diameter and with the properties necessary for compatibility.

Previously unforeseen opportunities for the production of artificial blood vessels are created by tissue engineering. For this, cultures of fibroblasts, smooth muscle cells and endothelial cells, which are taken from a small tissue biopsy from the respective patient, are settled onto a suitable biomatrix. First, a cell suspension is created by the proteolytic degradation of the ECM. For multiplication in culture, the cells are stimulated with growth factors. After that, a three-dimensional network of fibroblasts is grown in a flat biomatrix. This construct is formed into a small tube that is then settled by smooth muscle cells on the outside and by endothelia on the inside. Next, the construct has to mature and a medium is run through it during culture in order for it to adjust to the rheologic stress conditions of flowing blood. Constructs created can withstand an experimentally produced interior hydrostatic pressure of 1000 mmHg, which is six to eight times the natural systolic blood pressure. There is no doubt that in the future artificial vessels will be implanted during bypass operations after further optimization as a medical product. Furthermore, it is likely that not only coronary vessels, but also large diameter arteries and veins can be created using this principle.

[Search criteria: blood vessels tissue engineering]

4.2.7
Heart Valve Defects

Infections and chronic changes can lead to reduced function or to a stop in function of the heart valves. The heart surgeon must decide if a technical implant of metal or polymer materials, a human donor heart valve, or a xenograft from pig is to be implanted.

Years of experience in implantation of biological heart valves has shown that they can work perfectly for many years, but that they can lead to tremendous complications. Problems can be posed by the endothelial covering, atypical calcification within the connective tissue of the valve and accumulation of small gaseous bubbles in areas with reduced blood circulation.

To improve the creation of the endothelia on the surface of the valve and to avoid the atypical creation of products of calcification, recent biological heart valves have been improved through tissue engineering. After removal, the heart valves are incubated with proteases and detergents to remove all cellular particles from the tissue. What remains is the ECM of the heart valve, which is settled with autologous (from the patient) fibroblasts and endothelial cells. Afterwards, a longer period of culture follows to enable the settled cells to build mechanically stressable tissues, following which the constructs can be implanted.

[Search criteria: heart valves tissue engineering]

4.2.8
Neural Damage

Neural tissue does not renew its neurons with dendrites and axons for the entire lifespan of the adult organism. This missing ability of the neural tissue is especially problematic with lesions of the spinal cord. Here, the axonal connections to the muscles are interrupted, among with other problems. Consequently, paralysis occurs. Hope for regeneration is given by stem cells from the patient which are developed into neurons under culture conditions. It is necessary to use an artificial ECM like Hydrogel to get the neurons to grow the necessary dendrites and axons. The neurons are then implanted into the damaged segment of the grey matter in the spinal cord. The following phase of regeneration is especially difficult. Every axon of a future motor neuron now must find the right connection to neural fibers. Additionally, the axon must develop in the right direction over a distance ranging from many centimeters up to 1 m in order to reach the connection site on the respective muscle fiber.

Of special interest are experiments where stem cells are used in segments of the spinal cord with transverse lesions. The idea behind this technique is that stem cells develop into neurons, whose extensions grow into the tracts of the separated segments in the spinal cord, regenerating the interrupted muscle innervation.

[Search criteria: nerve repair tissue engineering]

microenvironment, the ability to regenerate is extinguished and the tissue starts to degenerate.

The functionality of stem cells is to be derived from the microenvironment of the niches. An internal mechanism of the niches tells the stem cells whether to divide in a symmetrical or an asymmetrical manner. Identical cells follow from symmetrical division, while precursor cells and later functional tissue cells follow from asymmetrical divisions. It is known that this development cannot be realized solely by the precursor cells, but that a neighborhood of differentiated cells is also required. Further development is controlled by signal molecules and/or cell–cell contacts. This means that the neighborhood as well as the individual gene program is required for the total development process. If the niche is experimentally destroyed, natural regeneration is also inhibited. Thus, in the case of a resettling of the niche, the regeneration starts again. Most of the known niches are located in the area of a basal membrane. Extracellular proteins of the matrix appear to cause a special microcompartalization, by means of which special adhesive properties appear to attract the stem cells. Furthermore, through the integration of signal molecules into the ECM, special morphogeneous programs could be saved.

All these properties presume the existence of genes that are capable of controlling, on the on hand, permanent self-renewal of stem cells and, on the other hand, the way into the development towards differentiation. Possibly these are piwi, Sox2 and Oct4. It is not yet known which other gene groups may come into question, how far the methylation of histones is part of this process and how these genes can be activated or suppressed in the right chronological order. Knowledge of this process will enable us to determine how stem cells develop and, thus, possibilities could arise to lead differentiated cells back into stem cells in order to generate new tissues from them. The importance of the surrounding environment for stem cells can clearly be seen here. Under optimized culture conditions, the redevelopment of oligodendrocytes into stem cell-like precursor cells with O2A could be shown.

Stem cells can multiply at will and differentiated cells can develop from them. This process can be found in various places in an embryo. The further the development proceeds, the fewer division processes take place and these are replaced by the proceeding differentiation. Within the adult organism, single stem cells remain which are obviously spread out in the tissues. However, there are no markers currently available to identify the individual cells and they remain invisible in most cases. In recent years, most of the knowledge of stem cell populations has been gained by examining the testicles, skin and intestine.

In order to generate functional tissues from proliferating cells in tissue engineering, they have to be settled on an artificial biomatrix. In most of the cases, scaffolds made of hydroxyapatite/tricalcium phosphate, polyglycol or polylactide acids are used. Additionally, growth factors such as BMP and optimized conditions of culture are applied. Most attention has, thus, been paid to the development of the tissue. There is still little knowledge of the whereabouts of the required stem cells. It is, however, solely this population which is of decisive importance for the long-term survival of the construct. This is why great efforts are taken in order to develop gene-activating (smart) matrices which improve the microenvironment of the stem cells. On the one hand, the stem cell population can be kept in one place and, on the other hand, the activities of development can be influenced by means of the release of morphogeneous signals. It becomes clear that, in this area, tissue engineering overlaps with genetic engineering. Hopes are that diseases such as epidermolysis bullosa can be healed by this combination. The dissolving of the epidermis of this disease is caused by a defect in the laminin molecule of the basal membrane which does not guarantee any functional anchoring for the cells in the epidermis. In future, it is possible that implanted stem cells could compensate for the genetic change of the dissolving of the basal membrane.

[Search criteria: stem cell self renewal proliferation symmetric]

5.2.8
Plasticity

Special stem cells have been described in the literature that have been isolated from adipose tissue of the adult human, for example. Adipose tissue can be obtained during almost every surgical operation and the stem cells isolated therefrom can be deposited in tissue banks.

In order to obtain stem cells, the adipose tissue is treated with proteases, and the stem cells are isolated and put in culture for multiplication. Following addition of dexamethasone, ascorbin phosphate and glycol phosphate to the culture medium, osteoblasts develop, while, after addition of insulin, TGFβ and ascorbin, phosphate chondroblasts can be seen (Tab. 5.3). Cells with muscle-specific abilities (myoblasts) should develop after addition of dexamethasone and hydrocortisone.

Tab. 5.3: Development of stem cells obtained from adult adipose tissue. No functional tissue cells develop, but precursor cells (blast cells) have initial tissue abilities.

Cells	Culture medium	Serum	Additional substances
Control group	DMEM	10% FCS	none
Fat-like cells (adipoblasts)	DMEM	10% FCS	isobutylmethylxanthine, dexamethasone, insulin indomethacin
Bone-like cells (osteoblasts)	DMEM	10% FCS	dexamethasone, ascorbin phosphate, glycol phosphate
Cartilage-like cells (chondroblasts)	DMEM	1% FCS	insulin, TGFβ, ascorbin phosphate
Muscle like cells (myoblasts)	DMEM	10% FCS/5% HS	dexamethasone, hydrocortisone

[Search criteria: stem cell plasticity differentiation tissue]

5.2.9
Diversity of Development

Analogous to embryonic development, the question of competence arises when stem cells are to develop into a functional tissue within an organism. Competence means the ability to respond to certain stimuli of development within a certain time range. It is known with stem cells that they can produce certain types of cells with factors like derivatives of retinoic acid, for example. If, however, a population treated with retinoic acid is cultivated as a whole, it is shown that only a certain percentage of the cells develop into the desired tissue. No description can be given as to what types of cells and tissues develop from the remaining part of the cells. In theory, they could, in part, remain stem cells and, in part, form completely different tissue cells. The question whether these cells had lost their competence for development, retained it or even gained further competence is problematic. With an intended implantation, the risk exists that different kinds of cell competence are included in heterogeneously composed constructs of this kind. It cannot be foreseen whether only the desired tissue develops or, in addition, a completely different tissue or possibly even a tumor, e.g. as can be observed with teratocarcinoma (Fig. 5.4).

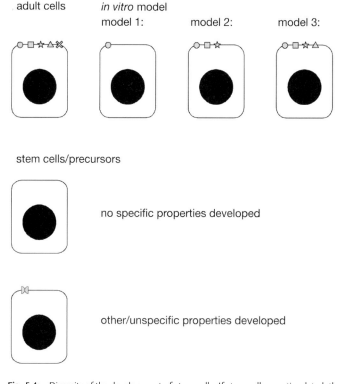

Fig. 5.4: Diversity of the development of stem cells. If stem cells are stimulated, they can develop different properties of tissue cells. However, part of the population does not react to these stimuli and another part possibly even develops atypical properties.

The question of competence of cells will be of central molecular biological importance for the future generation of tissue constructs. As there are not yet any clear concepts about competence in the development of basic tissues and their special forms, some important questions should be discussed here.

If the quality of the construct is to be improved, knowledge of the competence of cells will be analytically useful. One has to wonder whether tissues require different media in the early and terminal stages of development. It has to be analyzed if any applied hormones have the same effect in barely developed tissues as in the terminal phase of differentiation. It will have to be determined at which time the different cell populations require a morphogenic impulse in the various tissues and organs. It will thereby become clear that different cells develop at different speeds and require different morphogenic stimuli. Since one stimulus may harm another, attempts will be made to keep the duration of stimuli as short as possible. This will not be possible until more knowledge about the competence of our tissues becomes be available.

[Search criteria: stem cell tumor risk diversity]

5.2.10
Teratocarcinoma

Research on stem cells started in the mid-1970s. At that time, complex tumors were used which, apart from various differentiated cell types, also included a population of undifferentiated cells called EC (embryonic carcinoma) cells. This cell population was capable of developing early embryonic ectoderm, mesoderm and entoderm under suitable culture conditions. There are great concerns regarding the therapeutic use of these cells since they are aneuploid and are derived from tumors. Furthermore, it was known that teratocarcinomas could be experimentally produced if, for example, early embryos were subcutaneously implanted into mice. For this reason attempts were made to cultivate cell populations, not from tumors, but from early embryonic stages under various culture conditions. Cell lines with pluripotent development properties arose. How far these cells are still identical to the cells of the inner mass of the embryo is currently the subject matter of intensive research.

[Search criteria: embryonic stem cells teratocarcinoma potential]

5.2.11
Responsible Use of Stem Cells

It is possible to create cell lines by experimental multiplication of stem cells. In principle, during the work with cell lines of this type, problems that are not immediately obvious may occur. These includes contamination and cross-contamination as well as the risk of infections by animal cells. These problems need to be taken very seriously with the medical application of stem cells and, in particular, of stem cell lines.

Under optimal culture conditions, stem cells divide symmetrically when identical daughter cells are to be created. It is therefore possible to multiply cells at will prior to with limited development potential can be gained from the later genital loin region of the human embryo at the end of the second week. These cells can create a host of cell types, but the ability to create a whole germ is obviously lost. A compromise between scientific and ethical needs could be reached with stem cells from blood of the umbilical cord. These also are pluripotent stem cells, but, in comparison with embryonic stem cells, have lost development potential.

If, for example, it is tried to generate a tissue construct from stem cells, the procedure is purely empirical. A respective stem cell population is used and incubated with a factor that causes the cells to develop into a certain type of tissue cells. Then, the induced cells are brought into contact with a scaffold and the construct created is cultivated. After some time it can be determined whether the construct is more or less successful (Tab. 5.6).

When functional tissue is to be created from stem cells of precursor cells under *in vitro* conditions, critical questions on the final point of the construct required have to be asked. In this way the stage of differentiation the selected construct will reach and if or how long this differentiation is maintained should be clarified. However, there is no guarantee that stem cells develop in the same way as adult cells. In certain areas they can develop the same abilities, but at same time develop others suboptimally or not at all. Furthermore, it is possible that they will show totally atypical characteristics, as the development of embryonic cells is mostly controlled by totally different mechanisms than those maintaining the vital functions of adult tissues.

Numerous investigations on the regeneration of the intestine epithelium have been conducted *in vitro*. This showed, among others, that optimally preserved crypts can be isolated and taken into culture. The cells grow from the crypts and create a confluent monolayer on the culture dish. Hopes with this experiment were that during the creation of subcultures, the differentiated cells would be lost and the stem cells would keep proliferating. The results were that these cultures could not be subcultivated effectively and so the isolation of stem cells was unsuccessful. However, some cells produced a mucous-like secretion under such conditions. If these cells then were injected into a rodent subcutaneously, after some time it could be observed that some cells produced mucous and by this clearly developed into goblet cells. Furthermore, one could see that the implanted cells created crypt- or cyst-like structures under the skin. It is obvious that there are great differences between the environment of the culture dish and the subcutaneous environment. However, it was not possible using this method to generate a pure population of stem cells and from them, in turn, create a real functional tissue.

Autologous skin transplants are mostly carried out by cultivated keratinocytes being transplanted on a suitable artificial dermis. In most cases the constructs heal well, but problems often occur later on. As the keratinocytes have to be renewed permanently, the capacity of division of the stem cells included in the transplant can be overtaxed and thus be depleted too soon. Therefore, the long-term success of the transplantation will solely depend on the pool of the stem cells included. These have to remain during the isolation and the subsequent cultivation. The same applies to the recreation of a damaged cornea. In this case, cells from a hair follicle or from the limbus area of the cornea that include the required stem cells could be used therapeutically.

With the exception of hematopoetic stem cell therapy, most information about the use of embryonic stem cells has been gained from animal models. For example cardiomyocytes from embryonic stem cells of the mouse have been generated in culture and have been implanted into the hearts of mice. Implants have been created that promise long-term success. There are also promising steps with neural tissue. Damaged spinal cord of rats can be settled by embryonic stem cells of mice. Thereby, the generation of astrocytes, oligodendrocytes and neurons can be observed, and, in addition, motor ability is improved greatly. The list of possible uses of stem cells in tissue engineering in always increasing, ranging from epithelial surfaces to different kinds of connective tissue.

[Search criteria: stem cells tissue engineering applications clinical]

5.2.15
Possible Risks with the Use of Stem Cells

The impression is often created in the public and in the media that almost every kind of functional tissues is created automatically during the cultivation of stem cells. This is not right. Rather, stem cell precursors of tissue cells produce cells with initial tissue characteristics. Only time can tell if, by use of a suitable ECM in combination with optimal culture methods, fully differentiated tissues can develop. It is not yet clear how functional development of tissue can be fully controlled experimentally. Therefore, a lot of basic research has to be done in this area in the future.

Many ethical and cell biological problems remain with experimental work with stem cells, and these have to be resolved before the full potential can be seen. These include, on the one hand, the fact that cells in the form of suitable stable cell lines need to be produced in sufficient amounts and be available for everyone in cell banks. On the other hand, we need to clarify how functional tissue development can be controlled. This includes the very important question of whether stem cells used after application of a morphogenic development signal only create the required tissue or if part of the cells develops into another tissue or, in extreme cases, into tumor cells, too.

It has to be considered that the insertion of a tooth crown made of ceramic or the implantation of an artificial joint made of metal can be achieved with comparably little risk. Should they break or lack function, such implants can be completely removed. This is different with an implanted tissue. It interacts with and is bound-up by its surroundings. If unwanted tumor cells develop from the stem cells, the may under certain circumstances not be completely surgically removed with the inserted implant. In this case the cells of the tissue construct have to be equipped with a molecularly controllable suicide program that can be activated if needed and which eliminates the cells by activating apoptosis. The implanted cells could carry a gene that starts apoptosis in the cells after the application of a special drug and selectively eliminates them from the body.

still secrete proteins, e.g. insulin, into the surroundings. Originally the DEAE–dextran microcarrier was used, which supports the attachment and the cell division of primary cells as well as cell lines. With recent encapsulation techniques, alginates and agarose as well as synthetic polymers based on polyacryl and polyphosphates are now used.

During encapsulation of cells it can be seen that the cover acts like an optimally semipermeable membrane, and this supports the exchange of oxygen and nutrients, as well as not protecting the secretion of specific products for therapy. At the same time it has ensured that the capsule is not overgrown by fibroblasts, which would limit the possibilities of diffusion and the formation of microvascular structures.

[Search criteria: polymers scaffold tissue engineering]

5.4.2
Biodegradable Scaffolds

Biodegradable polymers have the ability to dissolve during culture with cells or after the implantation and thereby are replaced by newly synthesized tissue-specific ECM. During the degradation of the polymer the area of change area is permanently renewed by the cells. Scaffold materials that have been frequently used for decades are homo- or heteropolymers made of poly(L-lactate) (PLA), poly(glycolate) (PGA) and poly(lactate-co-glycolate) (PLGA).

The special feature of these support and scaffold materials is their degradability. After completion of their primary function of support and cell settlement, the matrix should be degraded by different mechanism like polymer dissolution, hydrolysis, enzymatic degradation and dissociation of polymer–polymer complexes. In optimal applications, the degradation products of the polymer are taken up into the biological circulation of the human body. The molecular weight of the degradation products should be as small as possible so that elimination by the usual methods is possible.

The biodegradable polymers used in tissue engineering, e.g. PLA and PGA, are aliphatic polyesters that belong to the poly(α-hydroxy)acids and can be produced by bacteria. The degradation is hydrolytic. The physical and chemical features in the production of PGA/PLA copolymers can be changed by variation of the lactide and glycolide proportions. The time of degradation also varies. With pure PGA fibers, the complete disintegration takes approximately 7 weeks; a PLA fiber only shows a roughly 10% lose of weight after 6 months.

If biodegradable polymers are used as scaffolding for the generation of artificial tissues, degradation metabolites like lactic or butyric acid are produced over time which are secreted into the medium of culture (Fig. 5.15). With this, two effects have to be considered. The scaffolds are no degraded equally, but from certain centers. The concentration of lactic and butyric acid is especially high at places where the monomers are released by degradation. Cells of the growing tissue located in this areas are exposed to an especially high local acidity, which is damaging to the cells and thus can influence the further development of the tissue. In this case, partial damage of the

Fig. 5.15: Degradation of scaffold materials. Biodegradable scaffolds release metabolites through the degradation processes that can damage surrounding cells through partial acidification of the medium (cross = dying cell). If increased degradation of the matrix takes place, the metabolites are released into the medium in higher concentrations. Thus, systematic damage of the whole construct can occur during culture, as well as after implantation.

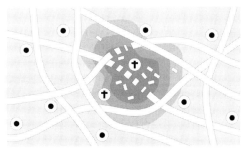

cultivated tissue is present due to local acidity, which can lead to central necrosis. If the metabolites reach the culture medium in ever-increasing concentrations, the physiological limit of tolerance is crossed at a certain time. In this case, systematic damage is done to the whole tissue. This occurs especially frequently if biodegradable scaffolds are used in the static environment of a culture dish. Therefore, we prefer perfusion cultures in which the developing tissue is continuously supplied with fresh medium. Here, the spent medium, i.e. that containing the degradation products, is removed continuously and not recirculated.

[Search criteria: tissue engineering material biodegradable]

5.4.3
Biological Scaffolds

A replacement for the damaged dura mater is necessary for patients with injuries of the skull and brain. Apart from different fleeces made of polymers, the dura mater of a deceased individual has proven to be an optimal matrix. Therefore, the dura of a specially selected deceased donor, whose medical history includes no signs of a possible risk of infection like AIDS or hepatitis, is removed. After preparation, the isolated dura is cut into patches of different size, packed, sterilized and deep-frozen. Parts of the dura mater can be used for multiple patients as required. It is obvious that no living cells are present in the dura with this way of preparation. The final material consists solely of mechanical stressable ECM and some remaining cell rests. After surgical implantation, fibroblasts of the patient migrate into the implanted dura from the sutures.

An analogous preparation, consisting of pure ECM, is the SIS (small intestine submucosa). The intestine of pigs is preferably used in its preparation. The mucosa with its lamina epithelialis, lamina propria and the lamina muscularis mucosae is mechanically removed from the lumen. The tunica muscularis on the outside of the intestine is removed. Cellular material is removed from the remaining submucosa by use of enzymes and detergents. It is then fixed, cut into shape, packed and sterilized. Such a piece of SIS can be sewn on surgically for covering stomach and intestine wall lesions. As it includes natural ECM proteins, the body's own cells migrate into it from the sutures of the implant and organize a new tissue network of epithelium, connective tissue or muscles, shortening the wound healing significantly. In some cases healing is made possible that would otherwise not have happened. Without the use of SIS in many cases the partial removal of the respective organ would be unavoidable. SIS is produced commercially and is available in different sizes for different surgical applications.

Dura mater matrices are tissue preparations that, apart from a very special ECM, also include dead cells and thus cell remnants. However, absolutely cell-free biological matrices can be produced by solving the cellular components from the tissue by biochemical extraction with detergents like Triton X-100 or desoxycholate, as well as with enzymes. In addition, different collagens can be isolated industrially from waste products of slaughterhouses, such as bones, skin, hooves, horns, swim bladders of fish and rooster combs, and these can be used in their purest form for the production of flat or three-dimensional scaffolds. In daily life, technically isolated collagens can be found as sausage skins, suture material, cooking and baking aids as well as capsules for drugs. Especially promising for tissue engineering, for example, are foams, which are produced as versatile collagen sprays.

The chitin armor of crustaceans provides scaffold material with impressive versatility. The natural polymer chitosan is closely related to cellulose, inhibits growth of bacteria and disables inflammations. It can store moisture, binds proteins and is biological degradable. A host of cells can be settled on chitosan scaffolds.

[Search criteria: tissue engineering biological scaffolds collagen]

5.5
Culture Methods for Tissue Engineering

Apart from cells and an optimal scaffold for the generation of tissues, suitable culture containers are required. While a large number of sterile single-use culture containers is available for the multiplication of cells, only a very limited selection exists for tissue growth at present.

The widespread introduction of disposable products in the area of cell cultures has led to a general trend of using ready-to-use and, thus, sterile packed standard products made of plastics since the beginning of the 1980s. Furthermore, clever marketing has contributed to the almost infinite confidence in the quality of cells that be cultivated in such culture containers. With this it became almost unthinkable that we can improve the culture environment, and thus improve cell and tissue quality. It is because of this that the methods of culture applied in the area of tissue engineering have to be given special attention

[Search criteria: tissue engineering bioreactor]

5.5.1
Petri dish

In the most simple case of tissue generation, a scaffold is created on the bottom of a culture dish. The cells are pipetted onto it together with the culture medium (Fig. 5.16).

Experimental data shows that, with good interaction with the biomaterial used, the cells settle within hours. However, with longer duration culture it is shown that the development into a functional tissue ceases at a certain time because the scaffold has contact to the bottom of the culture dish on one side.

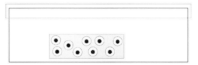

Fig. 5.16: Example of a scaffold settled with cells at the bottom of a culture dish. While the settlement of the scaffold with cells is feasible, more problems are posed with increasing culture duration of the construct due to the static environment. As scaffold lies on the bottom of the dish, insufficient nutrient supply occurs, especially at the borderline.

In the static environment of a culture dish this leads to stationary layers with a poor supply of nutrients and oxygen, which in turn naturally results in a negative influence on the tissue differentiation of the scaffold (Fig. 5.16). The longer the culture medium is not changed, the greater the risk for the construct. Damaging products primarily produced through the cell's metabolism accumulate very fast. In addition there are also metabolites created by the degradation of a biodegradable scaffold, for example.

In professional experiments on functional tissue maturation one has to rethink and consider that every tissue has special demands that require experimental adjustments (Fig. 5.17). Conventional single-use culture containers like Petri dishes only allow such a special modification of the tissue environment in the rarest cases. In a culture dish, cells can be multiplied almost at will, but this method is insufficient for different reasons for the generation of tissue. Methods of culture have to be applied which meet the physiological requirements of the single tissues and thus enable the development of specific abilities.

Adjustment of the tissue environment means that manual work has to be performed under sterile conditions. Scaffolds have to be selected, cut to the suitable size and adjusted to tissue carriers. After pipetting cells onto the scaffold, the developing tissue has to be cultivated under conditions as physiological as possible in special microreactors. Therefore, suitable tubes and connections are adjusted, and a source of the culture medium is installed. Finally, it has to be decided whether the tissue should be generated in a CO_2 incubator or under air. Depending on the selected strategy, a tissue-compatible buffer system is necessary to conduct experiments at a constant pH for weeks or even months. In contrast, experiments with proliferating cells can be conducted very quickly (within days in most of the cases).

[Search criteria: tissue engineering Petri dish]

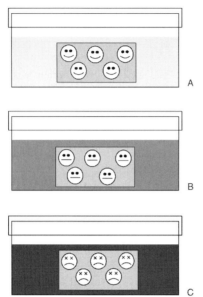

Fig. 5.17: The maturing tissue poisons itself in the static environment. In fresh culture medium the maturing tissue still finds good conditions (A). After a few hours the medium changes greatly due to the metabolism (B). After as soon as 1 day the medium can be so strongly accumulated with metabolites that further maturation of the tissue is prevented (C).

5.5.2
Spinner Bottles

Improved methods of tissue creation under *in vitro* conditions can, for example, be produced in glass containers with relatively high volume in which the culture medium can be kept in permanent motion by a magnetic stirrer (Fig. 5.18). In this way, the tissue construct, hanging on a thread, is exposed to a permanent stream of liquid.

Despite the greater volume of liquid, disadvantages of this method consist in the fact that the culture medium, like in a Petri dish, is not exchanged continuously and that

Fig. 5.18: Example for the culture of a tissue construct in a spinner bottle. The scaffold settled with cells is fixed to a thread. A magnet at the bottom of the container keeps the liquid in permanent rotation.

more and more products of metabolism accumulate with increasing culture duration. In addition, the cover of the container has to be opened at least partly so that the culture medium can be ventilated for oxygen supply and stabilization of pH. This means a higher risk of infection for a culture that lasts several weeks.

The advantage of the method is that due to the permanent movement of the medium, the tissue construct in not exposed to stationary layers and thus optimal removal of metabolic products metabolism from the interior of the constructs is possible.

[Search criteria: spinner bottle cell culture]

5.5.3
Rotating Bioreactor

Another possibility for improving the culture of tissue constructs is the rotating bioreactor (Fig. 5.19). This is a cylindrical chamber. A hollow space which houses the developing tissue construct and the culture medium is located in the interior of the cylinder is. A disk-shaped chamber is then fixed to the axis of an engine module. Consequently, the chamber along with the tissue construct can be moved up and down, and the tissue construct is subject to discontinuous microgravitation. This culture is also often conducted in a static environment, where no continuous supply of nutrients with new culture medium takes place. Released metabolites are not continuously removed and can damage the maturing tissue through their accumulation.

In a modified form of the rotating bioreactor it is possible to fill fresh culture medium into the chamber continuously or discontinuously and to remove the spent medium.

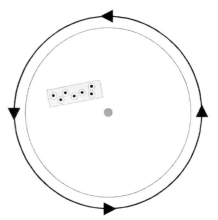

Fig. 5.19: Prevention of fouling and stationary layers at borderlines. The rotating bioreactor consists of a drum-like chamber connected to a propulsion axis. Culture medium is filled into the hollow of the chamber and the growing tissue construct is placed there. The construct is subject to discontinuous microgravitation as the chamber rotates.

[Search criteria: cell culture rotating bioreactor microgravity]

5.5.4
Hollow Fiber Module

With the exception of epithelium and cartilage, all other tissues require an intact capillary network for supply with nutrients and oxygen. Analogously, the supply of tissue constructs has to be secured. A capillary network for the cultivated cells or developing tissues can be simulated under *in vitro* conditions by a hollow fiber module (Fig. 5.20). With this it is even possible to place a single hollow fiber into a special culture container. At the same time the ends of the hollow fiber are connected to a thin tube, through which medium flows continuously with the help of a peristaltic pump. The cells or the developing tissue settle on the in- and outside of the hollow fiber. These modules now exist in different sizes and with different numbers of hollow fibers.

These modules have proved best in the creation of monoclonal antibodies. The hybridomas are kept on the outside of the hollow fiber, while the interior is streamed by new medium. Oxygen and low-molecular-weight nutrients are transported continuously, which pass the through wall of the hollow fiber by diffusion. However, the high-molecular-weight antibodies cannot pass through the fiber and therefore accumulate within the cell compartment, where they are harvested.

Hollow fibers can consist of totally different materials, e.g. polysulfone, acrylpolymers or cellulose acetate. Cells react very differently in hollow fiber modules dependent on the tissue. Apart from the synthesis of monoclonal antibodies and cytokines, effective production of growth hormones and insulin has been shown in such modules.

In a capillary module, new culture medium with nutrients and oxygen can be continuously supplied to the growing tissue, and, therefore, low-molecular-weight metabolites are not accumulated, but removed. If the developing tissue grows on the outside of the capillaries, very constant environmental parameters can be maintained by regular streaming of the medium.

With this one has to pay attention that the tissue layer should not be thicker than 150 µm and that the construct is near the hollow fiber. If a bundle of approximately 150–300 parallel hollow fibers with a diameter of roughly 250 µm is used, the space between the hollow fibers, i.e. the artificial interstitium, can theoretically be supplied without problems. However, practical application shows that the tissue does not spread in the module as well as desired. Places occur which are tightly overgrown by cells, while in other areas only single cells are present. Areas are often found that only consist of dead tissue. It is relatively difficult to analyze the reasons for this. In question are apoptotic triggers, necrotic lesions and also simple stationary layers, in which no continuous exchange of metabolites developed. In every case the analysis is very difficult and

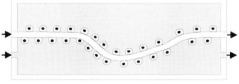

Fig. 5.20: A typical hollow fiber module consists of a case and the installed bundle of hollow fibers. Cells are settled of the inner and outer space of the hollow fibers. This concept ensures that the interior of the hollow fiber as well as the outer areas are streamed by media.

very extensive; single capillaries cannot be removed from the module, and, consequently, histological profiling can only be performed on the complete module and on different areas.

[Search criteria: tissue engineering culture hollow fiber]

5.5.5
Perfusion

When cells are cultivated under the static environment of a culture dish, they are subject to a certain volume of medium with the necessary nutrients and metabolites during the inoculation. However, over time and due to the increase in cell numbers, various parameters in the culture medium change continuously.

Static environmental conditions can have fatal results especially with growing tissues, because deeper cell layers cannot be supplied with sufficient nutrients and oxygen, and damaging metabolites are not adequately removed. A solution is the use of perfusion containers (Fig. 5.21) in which a constant environment exists as they are continuously evenly streamed with fresh medium.

Fig. 5.21: Culture of a tissue construct in the center of a perfusion container. The container is streamed evenly by fresh cell culture medium. As the construct rests on a tissue holder, it can be reached evenly by medium on all sides.

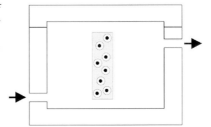

Maturing tissue constructs should not be put on the bottom of a perfusion container. This would lead to stationary layers between the tissue construct and the bottom of the reactor. Thus, tissue carriers are needed that fix the construct mechanically, placing in the interior of the container and ensuring that the medium reaches it evenly (Fig. 5.22).

Fig. 5.22: Examples of tissue carriers with different matrices for optimal adhesion of the cells.

Tab. 5.9: Guidelines for the simulation of a tissue typical environment. Cell multiplication, adhesion and tissue-typical differentiation are experimentally conducted in successive steps.

In vivo	*In vitro*	Method
Cell multiplication	culture dish	growth factor, serum
Adhesion	matrix, scaffold, biomaterial	matrix in the tissue carrier
Differentiation	perfusion container, hormones	adapted culture medium

enable the cells to develop tissue-typical differentiation, they must be settled on a suitable scaffold. In addition, it is necessary to use suitable culture media which, depending on the tissue type, include very different morphogens, growth factors or hormones.

[Search criteria: perfusion culture continuous exchange bioreactor]

5.6.1
Tissue Carriers

Cell of a maturing tissue require a suitable ECM as a foundation for optimal development, to which they can attach, and on which they can multiply and develop. Carrier systems are preferably used in order to not damage constructs and to be able to handle them well manually (Fig. 5.27). From the stage of cultivation over the period of differentiation to the experimental or clinical application, the constructs can be transferred with a forceps without problems. As both rigid matrices as well as flexible ones, made of totally different materials, are used, various suitable tissue carriers have to be available.

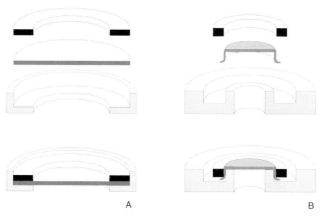

A   B

Fig. 5.27: Framework for tissue carriers for the uptake of rigid (A) and flexible (B) matrices. The respective matrices are mechanically fixed between a holding and a stretching ring.

Fig. 5.28: Example of a tissue carrier for the uptake of flexible matrices (diameter 6 mm). A thin collagen matrix can be recognized in the center, which is fixed like a ear-drum between the basis and the cover part of the carrier. The outer diameter of the tissue carrier is 14 mm.

Flexible matrices made of collagen are best placed in a carrier consisting of a special bottom and cover parts. The flexible matrix is stretched between both parts like the membrane of a drum (Figs 5.28 and 5.29). A tension ring is place upon it, which fixes the matrix.

Fig. 5.29: Example for a tissue of the embryonic kidney on a flexible collagen matrix in a tissue carrier.

The flexible matrix can be placed in the tissue carrier before or after settlement with cells. Furthermore, thin tissue preparations can be produced and fixed on the carrier with a tension ring.

Rigid matrices must be held place in different kinds of tissue carriers (Fig. 5.30). Therefore, a holding ring is used in which a great selection of natural or artificial ECM with diameters from a few millimeters to centimeters can be secured. For settlement, matrices like filters of polycarbonate or nitrocellulose as well as three-dimensional materials are used. The respective materials are placed in the holdings and fixed by a tension ring.

After that the tissue carriers are packed in foil and sterilized in a suitable container. The prepared carriers are best stored in a refrigerator over long periods of time.

[Search criteria: tissue carrier perfusion culture organotypic]

of using light, this method works via epifluorescence stimulation. Fluorescent dyes have to be used that intercalate in double-stranded DNA, for example, and so lead to a distinct dyeing of the nucleus in under fluorescence microscopy. Such dyes include propidium iodide, DAPI (4-6-diamidino-2phenylidol-di-hydrochloride) and bis-benzimide. This method is so sensitive that even single cells in a wide scaffold can be detected.

Cell detection with DAPI can be performed very quickly. The overgrown support is fixed for 10 min in ice-cold 70% ethanol, washed in PBS twice for 2 min, a DAPI solution (0.2–0.4 µg/ml) is pipetted on to it, incubated for 2 min in the dark and again washed in PBS two times for 2 min. The evaluation is performed by fluorescence microscopy under stimulation with UV light – the nuclei appear as radiant blue (Fig. 5.33).

Dyeing with DAPI and bisbenzimide is used mycoplasm tests. The suspected culture is first dyed with DAPI. If other diffuse fibrous dyeing occurs, apart from the nuclei, it is the dyed DNA of mycoplasm, which means that the culture is contaminated.

The settlement of cells on an optimal matrix takes only a few hours, while on a poor matrix no satisfactory attachment can be detected even after days. The method described is so sensitive that every single cell in a scaffold can be detected with it.

[Search criteria: cell detection DAPI support]

5.6.4
Perfusion Containers

After cell settlement, the tissue carriers can be placed in different kinds of perfusion containers. This permits continuous supply with fresh culture medium (Figs 5.34 and 5.35).

There are many other arguments in favor of creating artificial tissues in perfusion containers and not in a static environment. In a typical case, cultivated cells form a monolayer, while most tissues consist of several, even three-dimensional cell layers or partly thick layers of natural ECM or artificial biomaterial. The relatively thick layers have to be continuously supplied with nutrients and oxygen. Permanent streaming with culture medium can serve as a replacement for the missing system of blood vessels *in vitro* that maintain a constant nutrient supply. Perfusion containers permanently supplied with fresh culture medium and cleansed of the spent medium are best suited for this. With continual streaming of the culture, fresh medium nutrients as well as oxygen can be supplied to it, while at the same time metabolites of cellular origin that may harm the metabolism are removed. It is especially important that metabolites created by the biodegradation of biomaterials can be continuously eliminated. Furthermore, factors of cell differentiation (cytokines) that work in a paracrine fashion should always be kept at a constant level. Finally, the perfusion can be designed continuously or in defined pulses, which minimizes the generation of stationary layers between cells and biomaterials.

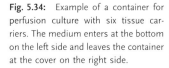

Fig. 5.34: Example of a container for perfusion culture with six tissue carriers. The medium enters at the bottom on the left side and leaves the container at the cover on the right side.

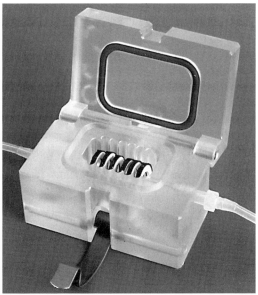

Another advantage is that, depending on the container used, the physiological environment for the developing tissue can be modulated, which is very similar to the natural conditions. In this way, the different tissues can be designed at smaller or larger scales. Above all, this modular technique enables one to discover on different cell biological levels how environmental influences have to be selected to act as optimal prerequisites for the generation of a functional tissue under *in vitro* conditions.

In perfusion culture (Fig. 5.36) the culture medium is transported by a peristalic pump from a storage bottle into a container into which one or more tissue carriers are placed. The culture medium used up by the cells is collected in a waste bottle and is not reused. In this way, the cultures are always supplied with nutrients and oxygen. Metabolism-harming metabolites cannot accumulate with this method.

The medium enters a perfusion container on the lower side of the container, spreads out over the bottom and ascends between the tissue carriers placed within to the top of the container, where it leaves again (Figs 5.34 and 5.35). The advantages of the construction are that the tissue carriers are streamed equally, that the container cannot run dry if culture medium is lacking in the storage bottle and that air bubbles generated are automatically removed. A host of tissues with unexpectedly high differentiation performance have been generated by this very simple method.

[Search criteria: perfusion culture container]

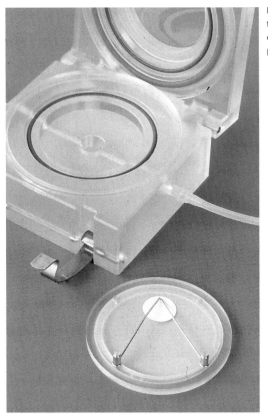

Fig. 5.35: A tissue engineering container used to create tissues with special three-dimensional surfaces under perfusion conditions.

5.6.5
Transport of Culture Media

Medium has to be transported continuously in perfusion cultures. This is best done by a peristaltic pump. The advantage is that the pump head with its several canals only has indirect contact with the culture medium through the wall of a silicon tube. In addition, because of the system of cassettes used, the respective sterile perfusion lines can be connected and disconnected without any risk of infection.

The transport rates are of importance. One should pay attention to the fact that the pump only transports small amounts of medium – this can be less than 1 ml/h. Furthermore, the pump should be controllable in such a way that it can work continuously as well as in pulses. This mode of pumping is required if stationary layers in the culture have to be avoided and scaffolds of, for example, large great material diameter are used. An interface should be present so that a cable connection to a personal computer can be established so that continuous rotation can be documented and individual programming of the pump can be performed.

[Search criteria: peristaltic pump culture medium]

Fig. 5.36: Principle of the perfusion culture. Medium is sucked from a storage bottle (left) by a peristaltic pump and is moved to a culture container in which tissue carriers are present. The spent culture medium is collected in a waste bottle (right).

5.6.6
Culture Temperature

Perfusion cultures can be performed in an incubator as well as at room air atmosphere on a warm plate. This should have a cleanable surface and generate a stable environment temperature of 37 °C. Such warm plates are mostly available in labs in which paraffin cuts have to be stretched. A cover made of plexiglass minimizes variations of temperature and pollution with dust, particularly if the tissues are to be generated over weeks or even months. If extremely stable (epicritical) temperatures have to be achieved, the perfusion container is put in a suitable water bath, as used for enzymatic tests, for the duration of culture.

[Search criteria: temperature cell culture perfusion]

5.6.7
Oxygen Supply

Culture media for the generation of tissues have to include sufficient oxygen to prevent cell death. There are two main possibilities for oxygenation of the medium. One is to lead oxygen into the sterile nutrient medium through a vent and an electronic control unit. However, this method has the disadvantages in long-term cultures that the injection of a gas easily causes contamination, the amount of gas has to portioned and the concentration of oxygen achieved in the medium has to be measured. All this is possible, but technically costly. Further problems are caused when not just one perfusion culture is used, but when many samples are being handled simultaneously. Finally, it should be considered that this method not only leads to accumulation of oxygen, but also to the formation of gas bubbles in the culture medium that can cause unexpected problems in perfusion cultures.

Another, simple method exists to keep the oxygen content of the perfusion medium at a constant level. A gas-permeable tube, preferably made of silicon, is used as the "lungs" of the perfusion culture. The tube should be as long as possible and have an inner diameter as small as possible. In addition, the wall thickness should be relatively thin. This generates a large surface for gas diffusion. If a silicon tube with an interior diameter of 1 mm and a wall thickness of 1 mm is led through room air atmosphere at 37 °C for the transport of culture medium, an O_2 balance is established between the

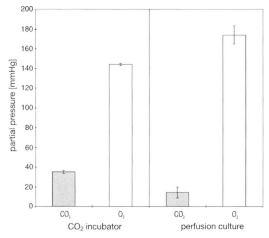

Fig. 5.37: Respiratory gases in culture media. Measurement of O_2 and CO_2 in IMDM in a CO_2 incubator and during perfusion culture in room air atmosphere. IMDM in the perfusion culture shows over 190 mmHg O_2 and so there is clearly more oxygen than the medium in the incubator.

culture medium and the surrounding room air atmosphere (Figs 5.37 and 5.38). For example, for IMDM at a pH of 7.4, 190 mmHg O_2 is measured by a gas analyzer if the medium can equilibrate against the atmospheric air in the silicon tubes during the transport process between the storage container and culture container. In contrast to this, clearly less oxygen is available to the growing tissue in the incubator at 149 mmHg O_2.

Measuring the oxygen can be performed surprisingly easily (Figs 5.37 and 5.38). The required device, i.e. an electrolyte analyzer, is present in clinical areas and in every emergency room. The measurement is performed in a silicon tube into which a T-junction is placed. After equilibration of the medium, a 1 ml injection is put to this port. Then, approximately 200 µl culture medium is slowly sucked up. To prevent gas diffusion and thus distortion of the measured data, the measurement of gases has to begin within 15 s. Therefore, the conus of the injection is connected to the suction needle of the analyzer. Depending on the type of machine, between 50 and 200 µl culture is are sucked up. Analytical data are shown after approximately 2 min. Apart from the O_2 and CO_2 content, the pH, respective electrolytes, and the concentration of glucose and lactate can be measured (Fig. 5.38).

[Search criteria: cell culture respiratory gases]

5.6.8
Constancy of pH

Not only the available oxygen, but also a constant pH is important for the generation of functional tissues. In our body the pH is, among others, regulated by the CO_2 solved in the body and the available $NaHCO_3$, and is kept constant at a very narrow physiological pH range between 7.3 and 7.4.

Depending on the organism, the sodium bicarbonate buffer system in the culture medium consists of $NaHCO_3$ and CO_2.

Fig. 5.38: Environmental parameters for tissue constructs. A print-out of a record of a gas analyzer after the measurement of a sample of IMDM is shown is. The values occur approximately 2 min after aspiration of the sample. The O_2 and CO_2 content of the medium, and the pH and the electrolyte values can be obtained from this, as well as the concentration of glucose and emerging lactate. Furthermore, the current osmolarity of the medium is also calculated.

```
     STAT PROFILE 9

25 Jan 01    11:12
.....................................
Proben Nr:5
Bediener Nr.:

Patienten Nr.:

Arterielle Probe
Probennahme:
FIO2:     20.9
BUNe:      0.      mg/dl
.....................................
Pat. Temp.    37.0 °C
pH            7.385
PCO2         41.3   mmHg
PO2          96.3   mmHg

Hk            6.    %
.....................................
Na+         140.8   mmol/L
K+            3.80  mmol/L
Cl-          98.5   mmol/L
Ca++          0.98  mmol/L
Glu         197.    mg/dl
Lak           2.7   mmol/L
.....................................
Hbc           1.9   g/dl
BE-ECF  -  0.4      mmol/L
BE-B    +  0.7      mmol/L
SBC          25.1   mmol/L
HCO3         24.9   mmol/L
TCO2         26.1   mmol/L
O2Sat        97.4   %
O2Ct          2.8   ml/dl
nCa++         0.97  mmol/L
An. Gap      21.
Osm         282.    mOsm
```

NaHCO₃ dissociates as:

$$NaHCO_3 + H_2O \Leftrightarrow Na^+ + HCO_3^- + H_2O$$
$$Na^+ + H_2CO_3 + OH^- \Leftrightarrow Na^+ + H_2O + CO_2 + OH^-$$

This reaction is dependent on the CO_2 partial pressure in the surrounding atmosphere. With low CO_2 partial pressure, the reaction equilibrium lies on the right side, which means the medium includes more OH and therefore is alkali. In order to prevent this, the incubator is gassed with CO_2 as required until the pH decreases to the required value. If the CO_2 concentration decreases, the pH value increases again and consequently CO_2 has to be added. pH stabilization in a CO_2 incubator is also based on this principle.

If a 5 % CO_2 concentration is offered in an incubator, then for every culture medium a corresponding amount of NaHCO₃ has to be added to achieve a pH of 7.4. If only 4 % CO_2 is offered, correspondingly less NaHCO₃ has to be added to achieve the same pH. For this reason the producers state the amount of NaHCO₃ and which CO_2 concentration has to be offered for every culture medium to achieve a constant pH between 7.3 and 7.4.

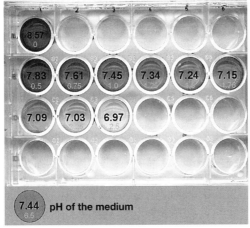

Fig. 5.39: Titration for the adjustment of the pH of culture media under atmospheric air. Aliquots of 1 ml of culture medium are pipetted into the depressions of a 24-well plate. Biological buffer substance is added to the samples in increasing amounts. With Buffer All, for example, a concentration-dependent series from 0.8 to 1.4 % is established. The samples are then equilibrated on a warm plate at 37 °C in room air overnight. Data show that the sample with a pH of 7.4 has to be detected by measurement and cannot solely be estimated on the basis phenol red coloration. In this example, 1 % Buffer All has to added to the culture medium to yield a constant pH of 7.4 under atmospheric air. Phenol red is an inaccurate color indicator in the area between 7.2 and 7.4.

If perfusion cultures are not conducted in an incubator, but under room air atmosphere, the pH value can be adjusted very easily if gas-permeable silicon tubes. In contrast to an incubator, the air naturally has a constant amount of CO_2. In the incubator, for example, 5 % CO_2 is available to the cultures, while in room air only approximately 0.3 % CO_2 is available. The experimental consequences can be easily foreseen. If a culture medium meant for a CO_2 incubator is exposed to room air, after a short time a discoloring of the phenol red to violet occurs, so that the alkaline and thus toxic area can be observed. This is solely caused by the amount of $NaHCO_3$ in the medium and the low content of CO_2 in the room air. Accordingly, the $NaHCO_3$ concentration in the medium has to be decreased to achieve a constant pH of 7.2–7.4 under room air. However, a constant pH cannot be adjusted by the decreased concentration of $NaHCO_3$ in the long run and, thus, a CO_2-independent buffer is additionally required for the stabilization of the pH. The addition of a biological buffer like HEPES or Buffer All (Sigma-Aldrich) has been proved to work best.

The correct pH for perfusion cultures under room air atmosphere has to be adjusted for every special medium (Fig. 5.39). Therefore, a culture medium should be used that has a lower content of $NaHCO_3$. For testing on a 24-well culture plate, 1 ml of culture medium is pipetted into every hollow. An increasing amount of biological buffer like HEPES or Buffer All is pipetted into each sample of culture medium. The 24-well plate is incubated over night on a warm plate at 37 °C under room air atmosphere. The pH of every sample is then measured in an electrolyte analyzer the next morning. The measured pH yields the necessary concentration of biological buffer that has to be added to the respective medium under room air atmosphere. Phenol red is an inaccurate color indicator in the range between 7.2 and 7.4. A constant pH in the culture medium in the perfusion culture can be maintained with this very simple method over any desired duration of time.

[Search criteria: cell culture acidosis alkalosis]

5.6.9
Starting the Perfusion Culture

At the beginning of a perfusion culture the cells are cultivated on a tissue carrier in a culture dish in a static environment for multiplication. One has to pay attention that the cells are attached optimally on the biomaterial used and later are not rinsed out by the continuous exchange of medium. After the assembly and sterilization of a perfusion line, the carriers are placed with in a container destined for use using forceps (Fig. 5.40). A clamp is placed on the silicon tube shortly before the container to prevent unwanted movement of the medium. To make the transition from the static environment to perfusion as optimal as possible, medium from which the tissue carriers has been taken is now pipetted into the container. After insertion of the pump tube into a cassette of the peristaltic pump the medium is transported at a rate of 1 ml/h into the container after the clamp has been opened. Over time the medium present in the container is exchanged with fresh culture medium. It is important for the good development of the growing tissue that a smooth, not abrupt, transition into the perfusion culture takes place.

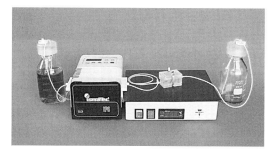

Fig. 5.40: Flow scheme of a perfusion culture under room air atmosphere. A peristaltic pump transports the medium from the storage bottle (left) into the culture container, which is located on a warm plate. The spent medium is collected in a waste bottle (right).

Dramatic cell biological changes result from the transition of a tissue from the static environment of a culture dish into the perfusion culture. Under the static environment of a culture dish the cells/tissues used are originally stimulated with serum or growth factor including culture media to attach to the available scaffold and to multiply as fast as possible. In contrast, at the start of the perfusion culture, the content of serum in the medium is reduced and, if possible, it is worked with complete-serum free medium for the next weeks. For the developing tissue this means that the proliferation activity, i.e.

Fig. 5.41: Work place for perfusion cultures is a T-shaped configuration. In the center is a cold box with the storage bottles for the culture medium. Three pumps suck the culture medium from here to the individual work lines.

quickened circulation from one cell division to the next, is stopped and so the possibility of a tissue-specific interphase is provided. As many functional tissue abilities as possible should be developed in this phase.

The perfusion cultures can be kept under a laboratory atmosphere as the lines are sterile and closed. The storage bottles are kept cool in a refrigerator (Fig. 5.41).

[Search criteria: perfusion culture continuous medium exchange conditions]

5.6.10
Gradient Container

In tissue engineering there is great interest in the creation of perfect skin equivalent, vessel implants, insulin-producing organoids, liver and kidney modules, as well as in the generation of urinary bladder, esophagus or trachea constructs. The biomedical application of these constructs will only be successful if the single epithelial cells have the necessary level of functional differentiation. They also have to be able to establish a close structural connection with the respective biomaterials used as artificial ECM in the creation of this constructs, for they alone yield the required mechanical stability. In addition, as the living tissue and the artificial matrix influence each other, it is important to examine how epithelial cells attach to the matrix, how this connection can be influenced experimentally and how long they can resist functional stress. This is mainly concerned with establishing the optimal tightness and transport abilities of the epithelial tissue. However, our current experimental knowledge on this is minimal.

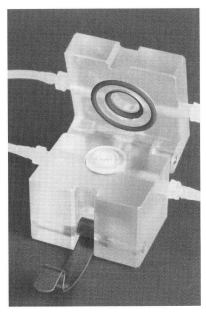

Fig. 5.42: A container for gradient culture with tissue carriers placed within. After closing the cover, the container can be streamed by different media above and below.

Epithelial tissues in an organism always form functional barriers and, thus, they are exposed to totally different environments on their luminal and basal side. Tissue carriers can be placed in a gradient container to simulate this tissue situation (Fig. 5.42). The tissue carrier divides the container into a luminal and basal compartment, which can be streamed separately with liquids, as under natural conditions.

The environment for epithelia can be simulated by tissue carriers into which a host of filters, foils or collagen membranes can be placed, and used as an artificial matrix for the settlement of the epithelial cells. The tissue carriers can then be placed in gradient culture containers which are divided into a luminal and basal compartments by the growing epithelium (Fig. 5.43). Research has shown that not only single growth factors, but influences like the surrounding ion environment have effects development determination.

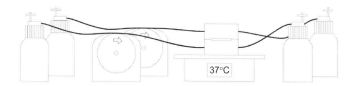

Fig. 5.43: Culture of epithelial tissue in a gradient. A tissue carrier is placed in a gradient container. The carrier divides the container into luminal and basal compartments that can be streamed by different medium, as under natural conditions.

What seems to be logical is often hard to realize experimentally. It is shown over and over again that cultivated epithelia do not develop their barrier function perfectly or that the function can be lost over a long culture duration of weeks (Fig. 5.44A). Non-tight epithelium (epithelial leak) is caused by insufficient confluence of the cells. A functional barrier cannot develop because of insufficient geometrical spreading or failure to seal to surrounding cells. In contrast, edge damage is caused by using suboptimal matrices for tissue carriers and/or differences in pressure or mechanical stress in the culture system. Edge damage always occurs at places where living cells, artificial matrix and tissue carriers are in contact, and thus are subject to great mechanical stress. Problems are also posed by the irregular occurrence of pressure differences between the luminal and the basal gradient compartment (Fig. 5.44B).

Fig. 5.44: Biological and technical problems in the culture of epithelia in a gradient. Epithelial tissue loses its barrier function due to insufficient contact to the holding tissue (A; edge damage) or due to a defective connection to the surrounding cell (B; leak).

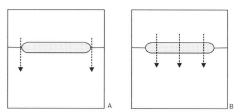

[Search criteria: gradient tissue perfusion culture]

5.6.11
Gas Bubbles

While gas bubbles are eliminated automatically and therefore do not matter in the streaming of a simple culture container, major unexpected cell biological and physiological problems appear with the use of gradient containers (Fig. 5.42). The accumulation of gas bubbles occurs easily if oxygen-rich culture media are transported by a pump. Favored places for their concentration are changes of material, e.g. where a plug in a connection of a tube has contact with the perfusion container. This effect can have fatal consequences for the epithelium growing within gradient containers. Air bubbles near tissues have to be prevented, because supply problems occur in these areas. The medium cannot spread evenly at the location of an air bubble. In addition, the joining of air bubbles leads to changes of the surface tension, which can cause surrounding cells to burst and, thus, greatly damage the growing tissue.

Perfect conditions for the culture of epithelia in a gradient container are provided if there are no pressure differences between the luminal and the basal compartments (Fig. 5.45A; $\Delta p = 0$). Due to the use of oxygen-rich culture media for culture, gas bubbles pose an unexpected problem. The reason for this is that medium is transported by the help of a peristalic pump through thin silicon tubes from a storage bottle to the gradient container. Accumulation of oxygen in the culture medium occurs by diffusion, as desired. This is necessary for the supply of oxygen to the tissue; however, the epithelium is subject to unexpected mechanical stress. During the transport of the medium the gas separates from the liquid phase of the culture medium. The occurrence of bubbles in the gradient container or within the tubes cannot be predicted. The gas bubbles remain at one place for a certain time and then they grow larger. The increase of size causes increasing congestion of the liquid and thus a change of the hydrostatic pressure, analogous to an embolus in a blood vessel.

The formation of gas bubbles takes place unpredictably either in the luminal or basal parts of the gradient culture. At first this leads to a still reversible arching of the tissue towards the compartment with lower pressure (Fig. 5.45B; $\Delta p > 0$). However, with

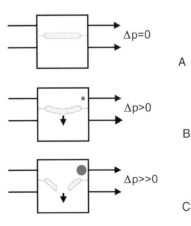

Fig. 5.45: Gas bubbles cause tissue-damaging hydrostatic differences of pressure and insufficient supply. If no gas bubbles are present in a gradient container, the same pressure conditions are in place at the luminal and basal sides (A). Gas bubbles can, for example, accumulate at the exit of a gradient container (B). The situation becomes fatal if in one compartment more gas bubbles (black point) can be found than in the other compartment. Consequently, pressure differences occur. This means that the epithelium placed within no longer can grow flat between both compartments, but arches forward to the side where the pressure is lower. If the hydrostatic pressure increases further, greater arching of the epithelium occurs. At an undeterminable time, the epithelium can no longer resist the pressure differences, rips occur and thus leakage (C).

Fig. 5.46: Minimizing of gas bubble formation during the transport of oxygen-containing culture medium. Culture medium is sucked from the bottle without contacting the cover (A). Gas bubbles can be eliminated in a gas expander module (B), in which the medium has to pass a barrier. During this, the gas bubbles separate from the liquid without changing the content of the solved oxygen.

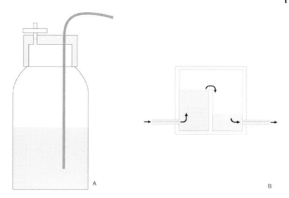

increasing pressure difference, the epithelium becomes physiological leaky and the tissues bursts (Fig. 5.45C; $\Delta p > 0$). Thus, the epithelium cannot form a functional barrier.

As gas bubbles occur where different polymer materials are in contact with each other, special covers for culture medium bottles have to be developed in order to minimize bubble formation. A silicon tube is led out of the bottle without the culture medium having to be in contact with the material of the cap (Fig. 5.46A). A gas expander module has been constructed to further decrease bubble formation (Fig. 5.46B). The medium that is pumped in reaches a small reservoir within the module, then has to pass a barrier and can leave the module again. Gas bubbles are separated at the liquid barrier of the module.

Measurements over several days showed that a clear reduction of gas bubble formation could be achieved by using optimized covers for the suction of culture medium and a gas expander module (Fig. 5.47B). In Fig. 5.47(B) it can be seen that fewer gas bubbles were registered with the use of suitable bottle covers and a gas expander module than without these additions (Fig. 5.47A).

For experiments with epithelia in a gradient container this means that damage to tissue can now be reduced tremendously by minimized gas bubble formation

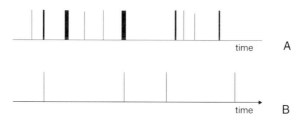

Fig. 5.47: Recording of gas bubbles by a detector over 96 h. During the pumping of culture medium many and some very large air bubbles usually occur (A). The number and size of the gas bubbles can be clearly reduced using the help of newly developed covers for the culture medium bottles and gas expander modules, without reducing the solved oxygen (B).

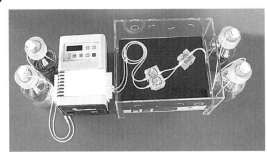

Fig. 5.48: Example for the culture of epithelia in a gradient container. A pump transports fresh culture medium from the both storage bottles (left) in the luminal and basal compartments of the gradient container. The spent medium is collected (right). A gas expander module is located before the gradient container to eliminate gas bubbles and, thus, prevent of pressure differences in the system.

(Fig. 5.48). Therefore, it is possible to generate more epithelia with intact barrier function.

[Search criteria: gas bubbles tissue culture perfusion culture]

5.6.12
Barrier Continuity

During culture for several days or weeks it is necessary to know in which environment the cultures develop. In order to do this, the environment is controlled by a blood gas analyzer. Through the T-junctions put into the tubes, 1 ml is sucked up with a sterile syringe. Measurements are made on both the luminal and the basal side, as well as before and after (Fig. 5.49). For culture under atmospheric air, the media are buffered by HEPES or Buffer All. A stable pH of 7.4 can be achieved during the whole duration of perfusion. Because of the low content of CO_2 in the air (0.3%), a relatively low content of $11-12$ mmHg CO_2 is measured before the container. In contrast to this, a high concentration of more than 190 mmHg oxygen can be detected, which is caused solely by the equilibration of the medium in the silicon tubes during the transport from the storage bottle into the container. The continuously high concentration of 415 mg/dl glucose shows that the exchange of medium is large enough so that a decrease of glucose does not limit aerobic physiological processes. Also, no unphysiological high amount of lactate can be observed, because the culture medium is collected in a waste bottle in the method described and not recirculated. Because of the continuous renewal of the medium, metabolites cannot have any damaging influence during the culture.

Epithelia in an organism develop in an environment in which they are exposed to the same liquid environment on the luminal and basal due to the lack of tightness that has still to be developed. However, physiological tightness is established during the process of polarization and establishment of tight junctions at the lateral cell borders. Therefore, the epithelia now can perform totally different functions on the luminal and basal side. A gradient is formed by which molecules can be transported through the epithelium very selectively.

If one transfers this natural development to the culture of a generated epithelium, one has to distinguish an embryonic environment from a functional environment.

Fig. 5.49: Example of measuring the physiological parameters during gradient culture with intact epithelium. Measurements are performed before and after the container. During the whole duration of culture the epithelia are subject to a gradient with salt stress (A) on the luminal side and standard medium (B) on the basal side (130 versus 117 mmol/l Na).

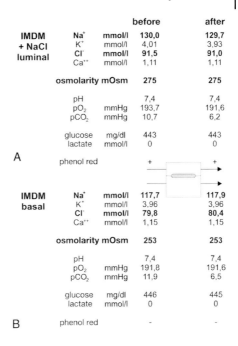

			before	after
IMDM + NaCl luminal	Na⁺	mmol/l	130,0	129,7
	K⁺	mmol/l	4,01	3,93
	Cl⁻	mmol/l	91,5	91,0
	Ca⁺⁺	mmol/l	1,11	1,11
	osmolarity mOsm		275	275
	pH		7,4	7,4
	pO₂	mmHg	193,7	191,6
	pCO₂	mmHg	10,7	6,2
	glucose	mg/dl	443	443
	lactate	mmol/l	0	0
A	phenol red		+	+

			before	after
IMDM basal	Na⁺	mmol/l	117,7	117,9
	K⁺	mmol/l	3,96	3,96
	Cl⁻	mmol/l	79,8	80,4
	Ca⁺⁺	mmol/l	1,15	1,15
	osmolarity mOsm		253	253
	pH		7,4	7,4
	pO₂	mmHg	191,8	191,6
	pCO₂	mmHg	11,9	6,5
	glucose	mg/dl	446	445
	lactate	mmol/l	0	0
B	phenol red		-	-

Embryonic conditions can be simulated in a gradient container if the same medium passes on both the luminal and basal side. This means that the epithelium is located in a permanent biological short circuit. This can be interrupted if different media are available on the luminal and basal sides. Formation of a functional polarization is supported by the generation of an initially small and then increasing liquid gradient. During a culture of an epithelium over several weeks in a gradient, it has ensured that the epithelium exercises its biological barrier function, that the gradient is maintained over the course of the culture and that it is not lost to surrounding influences (Fig. 5.50).

Therefore, the culture has to be examined continuously to determine if the barrier function of the epithelium is maintained. Optical control on the luminal side medium with phenol red and on the basal side medium without phenol red is used (Fig. 5.48). For further experiments, only such epithelia that maintained the liquid gradient dur-

Fig. 5.50: Example of measuring the luminal/basal gradient of sodium over 10 days. During the whole duration of culture the epithelia are subject to a gradient with salt stress on the luminal side and standard medium on the basal side (130 versus 117 mmol/l Na). The measured data show a constant gradient over the duration of culture.

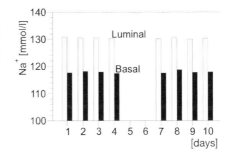

ing the complete culture duration and where no color mixture took place are used. In additional, analytic control with an electrolyte analyzer should follow (Fig. 5.50). In certain experiments the epithelia are exposed to a gradient with salt stress on the luminal side and a standard medium on the basal side over the whole culture duration (130 versus 117 mmol/l Na). Because of this, samples are taken on both the luminal and basal sides, as well as before and after the container. The maintenance of an intact epithelium barrier can also be recognized through the comparison of the sodium concentration as well as the measured difference of osmolarity between the luminal and the basal compartments.

[Search criteria: epithelia terminal differentiation barrier function]

6
Maturation of Tissue Constructs

In a natural tissue environment, each cell has its own individual cell biological inter-action with the surrounding ECM. Thus, when choosing material to create an artificial matrix, it is absolutely crucial to find out first whether the chosen biomaterial meets the requirements for the desired tissue construct in all respects.

It is relatively easy to observe the whole process when culturing individual cells, as the cells are mostly cultured in monolayers on flat matrix surfaces, which makes it easier for the cells to proliferate. Preference is given to transparent matrices, such as dishes made from polystyrene, glass, polycarbonate or aluminum oxide, which per-mit the observation of cell growth or, in neurons, the development of individual den-drites or axons under the microscope.

In contrast, experimental cultures of tissue and organs are often carried out at the interface of the culture medium and a specific gas mix in the incubator because cul-tured tissue proliferating at the bottom of a dish would hinder rather than help the experiment. Hence, the tissue fragments or primordial tissue are placed on a nitro-cellulose filter, perhaps even under layers of fibrin glue or agarose to prevent the tissue from floating and to stop uncontrolled growth. These examples show that the nitro-cellulose filter primarily serves as physical adhesion support for the cells and as a growth aid that is easy to handle.

However, when it comes to generating artificial three-dimensional tissue structures, the role of the matrix becomes crucial. When cells are grown on a scaffold, it is pri-marily the three-dimensional biomatrix that determines the size of the future struc-ture. If the construct is to grow larger, the scaffold must either be larger or the con-struct must produce its own matrix in order to grow.

Cells are placed on the chosen matrix and interact with it to form three-dimensional functional tissue. This will only work if the matrix not only supports the migration and adhesion of cells, but also furthers cell differentiation and the secretion of ECM pro-teins that confer physical stability to the construct.

Where biodegradable matrices are used, their task is merely temporary until the cells produce their own tissue-specific ECM and the artificial matrix degrades. This does not happen automatically and depends to a large extent on the cell biological support from the matrix used.

Cell biological interaction is a decisive factor when choosing a matrix for a specific type of tissue, and it must be said that predictions cannot be made for newly developed

Tissue Engineering. Essentials for Daily Laboratory Work W. W. Minuth, R. Strehl, K. Schumacher
Copyright © 2005 WILEY-VCH Verlag GmbH & Co. KGaA, Weinheim
ISBN: 3-527-31186-6

materials whether a particular cell type will thrive on them or not. The cell biological impact on various types of cells has to be reassessed for all newly developed biomaterial.

[Search criteria: tissue constructs differentiation maturation]

6.1
Primary and Secondary Contacts

In cell therapy, a suspension containing proliferating cells is injected into the area to be treated, whereas in tissue engineering, the cells must be established in a scaffold of ECM and grow into structured tissue that can be used as an implant. The matrix must therefore offer maximum support for adhesion and differentiation for the cells to grow into a tissue-specific structure. Additionally, it should provide a physically stable scaffold that permits natural three-dimensional growth. It would be better still if the matrix could biodegrade later to be replaced by tissue-specific material grown from the implanted cells. It is often assumed that cells that cooperate with the matrix will automatically produce functional tissue material. Unfortunately, this is not so. The matrix must have a number of specific properties in order to ensure the development of functional tissue. Apart from the cell biological properties, there are further essential requirements to be met to make the matrix biocompatible and to make sure the implant will have no toxic effect on the surrounding patient tissue.

As not only the growth, but also the functioning of a cell and thus the differentiation into tissue are largely affected by its physical surroundings, it is important to mimic the natural three-dimensional organization. The carrying structures used in the scaffold are supposed to provide a substrate that ensures optimal spatial and functional organization. As the cells must first adhere to the artificial matrix and then interact with it, the surfaces must possess a number of important properties. Wettable, hydrophilic surfaces are usually more suitable for adhesion than hydrophobic surfaces.

6.1.1
Adhesion

Cell adhesion is assessed through bringing a cell suspension in contact with the scaffold or biomaterial to be tested. After some time, the biomaterial is withdrawn from the cell suspension and kept in fresh culture medium for several days. In all likelihood, the material on which the cells have been growing contains some cells with stronger adhesion properties and others with weaker adhesion properties. The less well adhering cell population can be separated by centrifugation. The percentage of strong adhesion versus less strong adhesion among the cells can be established using fluorescent nuclear dyes in connection with light microscopy and electronic cytometry techniques. It is a fairly easy way to find out which biomaterial is, in principle, more suitable for cell adhesion than others.

Cell adhesion is a rapid indicator of the acceptance of biomaterial offered to cells (Fig. 6.1). However, this does not answer the next decisive question – do the cells settle evenly on the biomaterial and to what extent do they migrate into three-dimensional scaffold material? To find the answer, leave the cells to grow on various biomaterials for various lengths of time. Using fluorescent nuclear dyes and a confocal laser scanning microscope, the number of cells and their proliferation rate can be established. This is done using morphometry programs that show images of the three-dimensional distribution of cells in relation to the biomaterial substrate.

The quality of cell adhesion depends to a large extent on the biomaterial used. (Fig. 6.1), as can be demonstrated in cell cultures growing in Petri dishes. We know from experience how long a cell line or a primary culture takes to form a confluent cell layer on the polystyrene surface of a Petri dish. If the bottom of the dish is lined with an unsuitable polymer such as poly(2-hydroxyethyl methacrylate) the number of adherent cells declines so dramatically that there is no confluent monolayer. This example shows clearly the sensitive reaction of cells when exposed to an unsuitable surface.

When testing biomaterial, it is often found that the cells are not homogeneously distributed and aggregates have formed. While this may be a drawback for some kind of tissue growth, it may be an important prerequisite for the development of other tissue material such as gland tissue.

It is not clear why cells can settle on a variety of biomaterials that bear no molecular resemblance to their natural ECM (Fig. 6.2). It seems that a number of physicochemical surface parameters have an effect on adhesion, division and migration of cells.

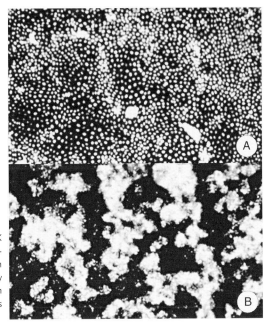

Fig. 6.1: Detection of nuclei in MDCK cells using propidium iodide on a suitable (A) and a less suitable (B) matrix. On the suitable matrix, the cells are evenly distributed over the surface, while on the unsuitable matrix, irregular clusters are found.

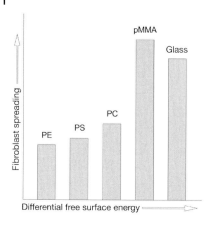

Fig. 6.2: Example of the differences in the growth of fibroblasts on PE, PS, PC, pMMA and glass surfaces.

Experiments with fibroblasts have shown that energy released by a surface has a distinct effect on the proliferation of cells growing on it.

It could be shown that there is a mutual reaction between a negatively charged cell membrane and the surface of the material used, and considerable differences in cell growth were found between PE, PS, PC, pMMA and glass surfaces (Fig. 6.2). On the basis of these results it should be possible, for example, to modify the charge of surfaces and thus improve cell adhesion. Through further chemical modifications, proteins such as fibronectin may be bonded to the surface of material to which the cells can connect selectively via anchor proteins. When using metallic substances, the difficulty is that their electric conductivity could induce redox reactions that could denature the proteins in the plasma membrane, damaging the cells irreversibly. Another crucial factor affecting tissue growth is the physical property of the surface and the pore size of the material used.

Experience so far with cells on biomatrices shows that there is no universally suitable material on which any kind of tissue would thrive. Instead, a very specific type of matrix must be chosen and optimized for each type of tissue. In other words, a matrix that is suitable for growing liver parenchymal cells is not necessarily the one on which to grow insulin-producing Langerhans islets and it is almost certain that it will be totally unsuitable for growing connective tissue cells.

Thus, if tissues are to thrive, the ECM has to have very specific properties. The – usually artificially produced – material must offer optimal adhesion points to cells. Only if the cells find tissue-specific anchoring can their integrin receptors pass on crucial information for further development inside the cell (Fig. 6.3). The required growth factors must either be contained in the structure or they must be added in order to encourage and maintain the differentiation of cells. In addition to cell receptors and ECM proteins, the role of matricellular proteins is very important as modulators that ensure the functioning of cells. Not only do they control ECM production, they also modify the effect of growth factors. This is done exclusively through receptor-mediated adhesion of the cell and the simultaneous expression of further receptors on the cell surface. These may, in turn, set off a whole range of other functions.

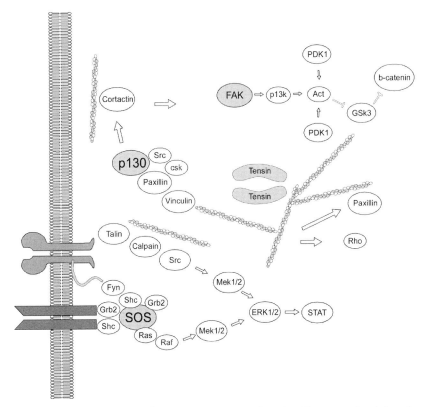

Fig. 6.3: ECM integrin signal transduction. Integrin receptors have the ability to transduce information from the extracellular to the intracellular space via the plasmic membrane. The information is passed on through cellular signaling cascades and used to regulate a wide range of functions.

When a new matrix is developed, no prediction can be made about its suitability for a certain cell type. This has to be established empirically in each case, as the differentiation profile of the cells in contact with the matrix determines its suitability. When we started out on this kind of work, we did not realize how sensitive cells can be in their reactions – ranging from desirable differentiation to undesirable dedifferentiation. Furthermore, it seems to be unclear why cells that settle on a scaffold do not automatically develop all the functional properties of a tissue, but remain more or less in a state of immaturity.

[Search criteria: cell matrix contact interaction primary]

6.1.2
Adherence

After cells have made primary contact with the scaffold, it remains to be seen if the cells emigrate or make permanent contact and develop tissue-specific properties (Fig. 6.4). From the mere fact that cells remain on a certain biomaterial it cannot be concluded that they are well anchored in the material, although this information may be vital, if, say, vascular prostheses are going to be optimized by establishing endothelial cells that can withstand a high degree of natural rheological stress in the bloodstream.

In order to optimize the adherence properties for vascular prostheses, biomaterials with growing cells are exposed to a variety of centrifugation cycles. During the process, cells may separate from the biomaterial to varying degrees. Control tests are carried out with blood vessel material.

Another way of testing the suitability of biomaterial is exposing the populated materials to various degrees of perfusion in a perfusion chamber. Care must be taken to keep the biomaterial with the cultured cells between two parallel chamber walls in order to produce a laminar flow. These experiments must be carried out under standardized conditions in order to obtain comparable results in different labs. They will deliver a picture of (a) the kinetics of cell adherence and detachment and (b) of cell rolling behavior on an ECM in a perfusion culture.

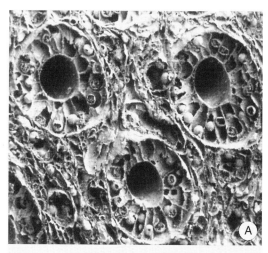

Fig. 6.4: Retaining the morphology of epithelial constructs. Microscopic view of collecting duct epithelium in a kidney (A) and cultured collecting duct epithelial cells on a natural substrate (B). In this case, the cultured epithelial cells are able to retain their characteristic morphological properties, which indicates that the collagen substrate is providing adequate biofunctional support for cultured renal collecting duct epithelium.

[Search criteria: cell matrix adhesion attachment]

6.1.3
Growth: ERK and MAP Kinases

The propagation of adherent cells is fairly easy under culture conditions. If they are given an optimal substrate, they adhere to it very quickly and remain on the substrate surface; if culturing conditions are good, the whole bottom of the dish will be covered with cells within a few days. A wide range of functionality stages can be observed – from various stages of cell division to the cell-typical interphase. What is remarkable about adherent cells is that they have permanent contact with the bottom of the culture dish throughout the stages of mitosis, cytokinesis and interphase. They never detach completely.

Cell division rates cannot only be stimulated by adding serum to the culture medium, but also by adding growth factor or via a change of the electrolyte content, which, in turn, affects osmolarity. These external stimuli must reach the inside of the cell to be processed on a cell biological level. ERK and MAP kinases are mediators of these functions and ensure the coordination of adhesion, adherence, mitosis and interphase within the cells (Tab. 6.1).

Once a cell has decided to adhere to a substrate, it will also make an individual decision whether it wants to remain in the functional interphase for a while or start mitosis. The signals controlling these processes must be transmitted and regulated between the inside and the outside of the cell. Information about the substrate reaches the inside of the cell mainly through mediation of the ERK system, and processes such as nucleotide synthesis, gene expression, protein synthesis and growth are stimulated as a consequence. These processes, in turn, are controlled by MAP kinases. Carbamyl phosphate synthase II (CPS II), for example, is a key enzyme for DNA or RNA synthesis. When epidermal growth factor was added, it could be shown that the ERK/MAP kinases transferred phosphate groups onto CPS II. This phosphorylation process can be speeded up by phosphoribosylphosphate (PRPP), which, in turn, enhances nucleotide synthesis and thus transcription activity. The signal from the cellular MAP kinases increases gene expression by activating rapid response genes. This is done through activation of transcription factors and phosphorylation of histone proteins and leads to changes in the molecular configuration – releasing DNA to be transcribed into mRNA. Via Mnk1 and eukaryotic translation factor eIF-4E, protein synthesis is activated simultaneously at the ER.

Tab. 6.1: Cell biological interaction of ERK (extracellular signal-regulated kinases) and MAP (mitogen-activated protein) kinases.

| Adhesion – Adherence – Mitosis – Interphase | | | |
Nucleotide synthesis	gene expression	protein synthesis	cell growth
CPS II	histone H5	Mnk1	cyclin D1
PRPP	access to DNA	EIF-4e	Cdk4/E2F
Enhancing nucleotide synthesis	enabling transcription	activation of ribosomes	activation of growth genes

6.4.3
Biophysical Factors

Functional differentiation in a culture system must be supported by exposing the cells to those biophysical parameters that will affect that particular tissue. These could be factors such as compression, rheological stress, sheering forces, temperature, partial gas pressures and many other factors.

Cartilage on the surface of a joint, for example, must withstand intermittent directional compression forces as they occur when using the joint. Compression provides a stimulus necessary to maintain cartilage differentiation. Cartilage in a joint that is kept still will become thinner, and changes in the structure and orientation of the cartilage matrix occur. Mimicking natural compression *in vitro* will help an artificial cartilage construct develop ECM with correct orientation. This can be achieved using a pneumatic unit or an eccentric disk exercising physiological and rhythmical pressure on the construct (Fig. 6.9).

A tendon is exposed to tensile stress that acts as a tensile stimulus on the tissue. The desired orientation and differentiation of cells in a tendon construct can be achieved by artificial tensile forces acting on the tissue *in vitro*. Cardiac muscle cells grown *in vitro* would be ideally grown on a flexible support so as not to restrain their rhythmical contractions. In a vascular construct, functional differentiation of endothelial cells can be enhanced by exposure to pulsating rheological stress. In the body, this is achieved by blood streaming through the vessels and it can be mimicked *in vitro* by rhythmic culture medium perfusion or keeping the cells in a laminar flow chamber.

Partial pressures of respiratory gases can have a decisive effect on the functional differentiation of a tissue. Tissues that are badly or not at all perfused with blood, such as cartilage, have a markedly lower oxygen content. In order to create optimal culturing conditions for such tissue, respiratory gas concentrations must be assimilated to the *in vivo* situation.

A tissue construct newly implanted into a patient will meet hypoxic conditions, as it is not yet vascularized and thus not supplied with blood. In a worst-case scenario, the implant is even encapsulated by fibrocytes, which further restricts the oxygen supply. Controlled lowering of the partial oxygen pressure *in vitro* might help to accustom the

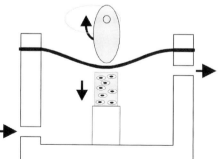

Fig. 6.9: Schematic representation of a tissue construct in a perfusion container with a rotating eccenter. The eccenter is there to expose the developing tissue to rhythmical compression.

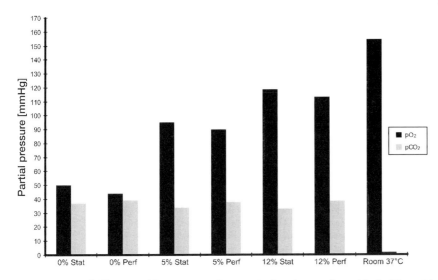

Fig. 6.10: Controlled lowering of the oxygen partial pressure in the culture medium, while the CO_2 partial pressure is kept constant. Measuring results of stationary (Stat) and perfusion cultures (Perf).

construct to hypoxic condition while it is still being cultured. This could help it survive the critical period after implantation (Fig. 6.10).

[Search criteria: compression stress tissue constructs]

6.4.4
Darling Culture Medium

Often, when you ask why a lab tends to use a certain culture medium rather than another for the current experiment, the answer will be that this is what is required in the protocol and this is what the predecessors always used. It would be inconceivable to use a different culture medium in this experiment.

In our research area, too, we did not think twice about using a medium we had been using for a long time – until the day we decided to investigate whether media that differ in their electrolyte make-up would elicit a difference in differentiation behavior. Artificially generated renal collecting duct epithelia were kept in a perfusion culture for 14 days, using media with varying NaCl concentrations under totally serum-free conditions. An immunohistochemical check was then carried out to determine the degree of differentiation that had been achieved (Fig. 6.11).

The results showed that all culture media produced epithelia with apparently perfect morphology. From the morphology aspect, it was impossible to link the epithelia to certain specific media, i.e. the tissues looked very much like each other. The immunohistochemical differentiation profile,in contrast, showed dramatic differences, using the two kidney-specific markers monoclonal antibodies (mAb) 703 and 503 as parameters (Fig. 6.11). The properties enhanced in cultures grown in media

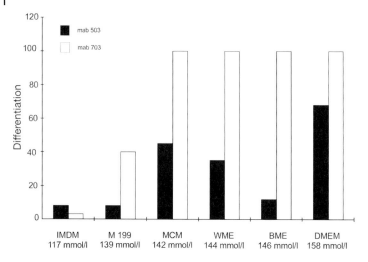

Fig. 6.11: Differentiation profiles of renal collecting duct epithelial cells that were grown in media with low and high NaCl content in a perfusion culture for 14 days.

with a low NaCl content differed from those grown in media with a high NaCl content. While mAb 703 binding increases in parallel with the rising NaCl content, mAb 503 shows no such correlation with the NaCl content of the medium. However, each of the culture media produces a very specific differentiation profile in the epithelium.

One must be cautious when analyzing the result of this trial series. Culture media are complex solutions. Although the difference in the culture media that had been used lay in their NaCl content, the differences found in the differentiation profile could be attributed not to a change in electrolytes alone, but to a variety of other factors. Nevertheless, our findings show that choosing a different culture medium can be expected to have repercussions on the differentiation profile of the tissue – resulting in typical as well as atypical tissue structures, due to excessive or reduced expression. The change could even lead to the production of proteins that are foreign to the tissue. Thus, neuronal cultures, when exposed to high levels of NaCl, were found to express muscle-specific proteins that would not normally develop within this tissue in the organism.

[Search criteria: culture media quality composition]

6.4.5
NaCl and Plasticity

Culture media are generally complex solutions with varying electrolyte concentrations and it must be established whether a single change in the NaCl concentration in the medium leads to a change in the differentiation profile of cultured kidney epithelium.

To that purpose, trial series were carried out with IMDM and increasing NaCl concentrations (Fig. 6.12). The rather unexpected outcome was that already small changes

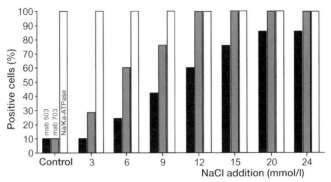

Fig. 6.12: Modulation of differentiation properties using electrolytes. Examples of renal collecting duct epithelium are shown in media of varying NaCl concentrations after 14 days of perfusion culture.

in the NaCl concentration (around ±6 mmol/l) can modulate differentiation properties and that increasing the NaCl concentration triggers a concentration-dependent reaction of the cells regarding markers mAb 703 and 503. In contrast, the production of a protein involved in the transport of electrolytes (Na/K-ATPase) is not affected by electrolytic changes.

[Search criteria: sodium chloride cell plasticity]

6.4.6
Natural Interstices

One of the unsolved problems in tissue engineering is the question how functional tissue develops from embryonic structures and, above all, what role environmental factors play in the process. These include, primarily, ECM and, secondarily, the electrolyte environment

Embryonic epithelium usually develops in a milieu where it is exposed to the same fluid milieu on its luminal as on its basal side. This changes with the onset of polarization, when lateral interstitial spaces are sealed off. The epithelium now acts as a biological barrier, featuring specific transport properties. As the luminal and basal fluid environments are now diverging, a gradient develops. Such developmental processes can be simulated in a gradient culture container (Fig. 5.42).

In such a container, the conditions for an embryonic epithelium can be recreated by perfusing the gradient container with the same culture medium. Conditions for adult epithelium can be created by pumping different media along the luminal and the basal side. Thus, the epithelium is on a gradient and can function as a valve that transports substances from A (luminal) to B (basal) or *vice versa* (Figs 5.42 and 5.43). This arrangement lends itself to investigating to what extent growing epithelium reacts to changes in its environment under near natural conditions.

Collecting duct epithelium from a mammal kidney serves as model tissue in our studies. It has two specific characteristics. First, it is generated from embryonic cells that derive from the collecting duct ampulla of a developing kidney – so it is actually a stem cell population. In a first step, these cells induce the development of all nephrons

development profile of cytokeratins in liver cells. Only through the experiments described above did we become aware of the major impact the electrolyte milieu can have on the differentiation behavior of an epithelium. This was achieved by just slightly modifying the NaCl concentration in the culture medium.

If electrolytes such as NaCl can induce the development of properties in embryonic tissue, it seems reasonable to investigate whether these acquired properties can be downregulated by withdrawing the stimulus. To that end, collecting duct epithelia were cultured in a NaCl gradient. After 14 days, standard medium was perfused again on the luminal and the basal side (Fig. 6.13B). The culture was then continued to day 19. It turned out that after the gradient had been removed, some of the properties, such as mAb 703 or CD9 binding, were maintained in the epithelium, whereas mAb 503 reactions fell to less than 10% of the cells. It can thus be concluded that the electrolyte milieu is not only crucial for the development, but also for the maintenance of protein expression in the generated epithelium.

[Search criteria: modulation differentiation gradient perfusion culture]

6.5
Step by Step

During the culture experiments just described using varying loads of NaCl, it struck us that the tissue-typical properties did not develop in the first few days, but only reached their full potential after 2 weeks at the earliest (Fig. 6.13). This must have a reason, but it took us a long time to find it. When looking at the morphology of the tissues, we eventually noticed that their volume had not visibly increased during culture. They had kept their original size. This finding led us to examine cell proliferation during culture, using immunohistochemical markers (Fig. 6.14).

The epithelial tissue was generated in a serum-containing culture medium where it stayed for 24 h. It was then transferred to a serum-free medium and cultured for another 14 days. At regular intervals, the constructs were immunohistochemically investigated, using antibodies to proteins such as Ki67 or MIB1, which help keeping track of the ongoing cell cycle (Fig. 6.14). While at the beginning of the culture mitoses were still frequent, as could be shown using the monoclonal antibody against cell cycle protein MIB1, no more dividing cells could be detected after the first week. The tissue had reached a postmitotic stage, as could be found in an adult kidney.

There must be some connection between the development of tissue properties and the simultaneous decline in the frequency of mitosis. While culture experiments in a gradient container showed that changes in the electrolyte milieu elicit changes in tissue properties (Fig. 6.14B), reactions in the tissue were it was remarkable that tissue reaction came very late, and only after the first week does the upregulation of properties set in. A second important event also takes place at that moment. When epithelia are changed over from a serum-containing to a serum-free culture, mitosis also stops at the end of the first week only (Fig. 6.14D). This suggests that cell-specific properties are only upregulated once mitotic activity of the cells has come to a halt.

Fig. 6.14: Regulation of the interphase in tissue constructs. The image shows the mitotic frequency in collecting duct epithelium immediately after serum has been removed from the culture (A) and after several days of serum-free conditions (B–D). While frequent mitosis could be shown at the beginning of the culture, no more dividing cells could be found after 2 weeks using the mAb to cell cycle protein MIB1. The tissue has reached a postmitotic stage as found in an adult kidney. It is precisely from the onset of this stage that the expression of specific characteristics can be observed.

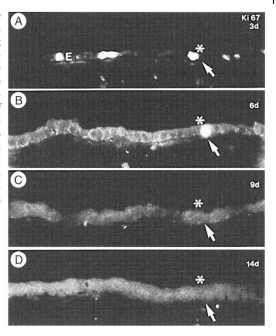

Data found in literature maintain that in the development of skeletal muscle, the cessation of mitosis coincides with the development of specific characteristics. This development is activated by transcription factor Pax3, which, in turn, induces two muscle-specific transcription factors Myf5 and MyoD. Both are members of a group called myogenic basic helix–loop–helix (bHLH) proteins, which activate specific genes by binding to DNA. MyoD activates the synthesis of muscle-specific creatine phosphokinase and acetylcholine receptor. As a result of this induction, MyoD is always produced in large quantities in order to bind to the DNA and keep up the activity of the gene. Thus, myoblasts come into existence.

However, muscle tissue can only develop if this is followed by another step – the fusion of the mononuclear myoblasts into polynuclear and, at a later stage, transverse striated muscle fibers. As soon as the myoblasts have stopped dividing, fusion sets in by myoblasts secreting large amounts of fibronectin, thus enhancing the synthesis of their fibronectin anchor protein ($\alpha5,\beta1$-integrin). The binding between integrin and fibronectin is crucial for all further development. If this binding process is blocked experimentally, e.g. by using an antibody, the next step in muscle development, where chains of myoblasts congregate, cannot follow. Glycoproteins such as cadherins and CAMs are also involved in the congregation process. Fusion can only take place if myoblasts recognize each other. Ca^{2+}ions seem to play a major part in this, as the fusion process can be activated by an ionophore such as A23187. In addition, metalloproteinases from the meltrin family also seem to play an important part.

When culturing tissue, it is essential to analyze very carefully if the developmental stage achieved under culture conditions matches the intended developmental stage.

While tissue in the growing phase will always contain a higher rate of mitotic cells, mitosis is of minor importance in phases of functional maturation. These natural developmental and physiological facts must be taken into account when adapting growing conditions to the needs of maturing tissue. Growing tissue is therefore cultured in media that contain serum or growth factors, while tissues that are not meant to grow in size, but to develop functional properties, serum and growth factors are not used, and the media should be electrolyte adapted.

[Search criteria: cell proliferation differentiation growth arrest]

6.6
Tissue Functions after Implantation

In order to produce a cartilage construct with high physical stability in the shortest possible time (Fig. 6.15), it must meet optimal conditions already *in vitro*. In most cases, isolated chondrocytes are cultured in DMEM/F12 media, although it is not known if differentiated human hyaline cartilage tissue can be kept alive over a longer period.

In order to find out, healthy human joint cartilage explants were cultured over periods of 2, 4, 6 and 8 weeks in a continuous medium perfusion system, using serum-free DMEM/F12. As a control, cartilage explants were cultured under stationary conditions in a culture dish over the same length of time, and then examined using histochemical, immunohistochemical and morphological methods. Their vitality was examined as well as their ability to maintain specific properties as a result of tissue

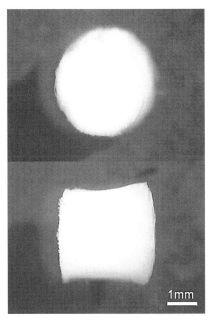

1mm

Fig. 6.15: Example of a native cartilage explant ready for perfusion culture.

differentiation. The results show that it is possible to retain differentiated cartilage tissue for a short time under all culture conditions tested. In the long term, however, perfusion culture was clearly superior. For example, in a perfusion culture, many specific differentiation characteristics were retained at a high level, whereas under stationary conditions, a high degree of dedifferentiation could be observed, resulting even in morphological changes in the explant.

[Search criteria: maintenance function tissue engineering differentiation]

6.7
The Three Steps of Tissue Development

The processes that make tissue structures develop in an organism are regulated on several cell biological levels (Tab. 6.3). As current literature in the field of tissue engineering shows, the chosen approach for the development of tissue constructs is often too simplistic. Only slightly caricatured, such protocols would read: "Take cultured cells, let them make contact with an artificial ECM and culture them in a dish in serum-containing culture medium". A critical overview of publications clearly shows that – depending on the matrix used and culture conditions – the differences to live tissues outweigh the similarities in most developing constructs. Morphological, physiological and biochemical characteristics have been substantially changed by cellular dedifferentiation. Furthermore, often, atypical proteins are expressed that could – if the tissue were implanted – lead to inflammation and rejection. As far as we know at this stage, it has not been possible yet to create tissue *in vitro* that equaled living tissue in quality.

Only now are we beginning to see a slowly growing awareness of the multiple cell biological and technical problems involved in the production of artificial tissue, and, more specifically, of obtaining adequate quality constructs. All procedures discussed so far are the results of long years of experience in an area that has only been little researched. As there was so little information available, our work focused on how tissue develops inside an organism. Many of the insights gained could be directly transferred from nature to the construction of tissue.

Although many of the cell biological and technical problems encountered during the maturation of functional tissue have been raised, but not solved, it has become clear that the first step is to learn to control tissue-specific proliferation and the various lengths of the interphase experimentally. Decisions on how long cells should be allowed to divide and at what point in time tissue-specific properties should be induced or maintained under culture conditions should be made on a case by case basis. This should be done with the natural life cycle of a cell in mind (Fig. 6.5 and Tab. 6.3). Thus, a cell cannot be expected to be functionally differentiated during mitosis. It is also time to say farewell to the cherished idea that cells and tissue should always be cultured in a serum-containing medium. The approach we use in our cell cultures is closely modeled on the natural development of tissues and involves three consecutive steps (Tab. 6.3). In the first step, cells must proliferate in order to provide sufficient material for

Tab. 6.3: Regulation of tissue development. During the generation process of tissue, the mitotic and the differentiation phase in the cell cycle must be clearly separated in the setup of the experiment, as these phases do not occur simultaneously, but successively in the natural cycle.

	Step 1	Step 2	Step 3
Objective	expansion of cells	onset of differentiation	maintaining differentiation
Epithelia			
Connective tissue			
Culture method	stationary	perfusion culture	perfusion culture
Medium	growth factors, FCS in medium	serum-free media	electrolyte-adapted serum-free media
Biophysical effects	none	some	enhanced
Hormonal stimulation	none	yes	yes
Reaction	rapid mitotic cycle	decelerated mitotic cycle	postmitotic interphase
Mitotic stress	high	low	low
Differentiation	low	increasing	high

the subsequent stages. The culture medium used contains growth factors, FCS or adult human serum. In contrast, the media used in the second step contain very little or no serum at all, because the aim is to lower the mitotic rate in the cells while inducing an upregulation of specific properties in the maturing tissue. In this culturing phase, the tissues are no longer kept in a stationary milieu in a culture dish, but in perfusion containers that permit continuous perfusion with fresh medium. Where serum seems to be indispensable, FCS should be replaced by low concentrations of depleted serum from an adult human donor. The experiments carried out so far showed that this stage takes at least 2 weeks. A third step follows to ensure that differentiation is maintained throughout the remaining culture process. Properties that have been induced and developed during the process must not be lost. However, we have to recognize that all artificial media that are currently available to us have only little in common with interstitial fluid in live tissue and it seems that we will have to live with compromises in the laboratory for a while yet.

[Search criteria: cell culture technique differentiation organogenesis]

7
Development of the Perfusion System Tissue Factory

While the proliferation of isolated cells in Petri dishes usually occurs without major difficulties, many morphological, physiological and biochemical differences are often found in maturing tissue constructs. This dedifferentiation does not occur in this form in native tissue. Numerous studies have shown that the quality of the artificially generated tissue depends to a high degree on the quality of scaffolds used, cell adhesion, intracellular communication and the respective culture conditions. All these factors have to be highly complementary otherwise tissue with undesired properties will occur. Typical examples are the expression of atypical collagen in artificial cartilage and bone constructs, hardening of artificially produced heart valves, loss of essential endothelial cell layers in vascular constructs, as well as down-regulation of specific cell performance, as observed in liver, pancreas and kidney constructs.

Tissues develop into highly functional structures in an organism. In the culture of tissue constructs under *in vitro* conditions, however, significant variations from this developmental scheme are observed. This shows that the basic culture conditions need to be improved. For this reason, we developed a modular perfusion system, called a "tissue factory". Due to the microenvironment that has been adapted to the special needs of the desired tissue, any kind of three-dimensional tissue can be generated in culture flasks. In order to create an improved physiological environment and, subsequently, a typical tissue differentiation, space for the growing tissue as well as for the artificial interstitium is provided.

A simple construction principle allows the production of tissue culture flasks as well as gas accumulation and gas distribution modules. These components may be used individually or in a combined module. It is conceivable that the above will form a standard biomedical method for the production of artificial tissue.

[Search criteria: tissue factory]

Tissue Engineering. Essentials for Daily Laboratory Work W. W. Minuth, R. Strehl, K. Schumacher
Copyright © 2005 WILEY-VCH Verlag GmbH & Co. KGaA, Weinheim
ISBN: 3-527-31186-6

7.1
Requirements of the Culture System

Experience shows that functional tissues cannot be optimally manufactured in the static environment of a Petri dish. This is due to the fact that the generation of tissue structure is not only dependent on the action of a single growth factor, but on a multitude of cellular and extracellular factors. Therefore, culture systems are needed that simulate special environments of individual tissues. This will result in an optimal degree of cellular differentiation. In all known perfused culture flasks there is an unnecessary large void volume between the flask wall and the tissue that is filled with media (Fig. 7.1). During perfusion, this space acts with significant hydraulic force and mediates pressure changes, undampened, to the neighboring tissue. In addition, air bubbles arriving with the media finding a space in this area that increases in the course of the culture. This results in areas that are not sufficiently supplied with media due to the air bubbles. Furthermore, aggregating air bubbles produce surface tensions that cause mechanical damage to the growing tissue.

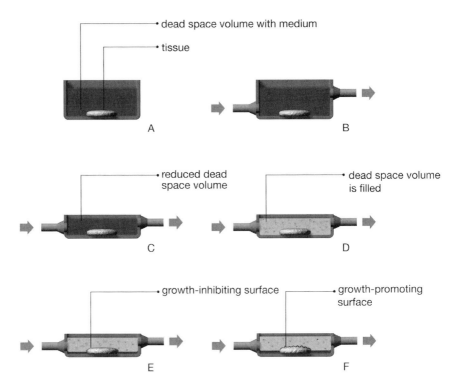

Fig. 7.1: Tissue factory for the production of tissue constructs – from the culture dish to the microreactor. (A) Tissue on the surface of a culture dish with a lot of void volume. (B) Tissue on the bottom of a perfusion container with a lot of void volume. (C) Reduction of the lateral wall height of the container to reduce the void volume. (D) Insertion of an artificial interstitium between the container wall and construct. (E) Growth-promoting artificial interstitium. (F) Growth-limiting artificial interstitium.

7.2
Artificial Interstitium

Optimization of the culture conditions at the boundary of developing tissues in the perfusion culture flask must provide pressure reductions and even pressure distribution in the medium. Dissolved gases need to be transported to and from the tissue in a continuously distributed way to avoid the formation of gas bubbles.

It also should be possible to experimentally modulate the growth-promoting or -inhibiting properties of the cultivated tissue. As a technical solution, newly constructed culture flasks are used that contain porous, biocompatible material with a quasi-capillary effect for pressure reduction and distribution, as well as for mechanical protection. At the same time, the void volume of the chamber is minimized.

Fleeces made from cellulose, glass fiber or sponge-like materials made from plastics, acellularized biomatrices or polymer matrices with cell biological functions are considered as fill material for the capillary space. This material can be distant or in direct contact with the tissue. The surface of the interstitium can be adapted individually and thereby influence growth or differentiation. It is conceivable that the artificial capillary space can be used to couple morphogens, growth factors or hormones similar to a natural ECM. In the course of matrix degradation these substances could be eventually released in the ultimate proximity of the maturing tissue and thereby contribute to the functional maturation of the cultivated tissue. The artificial capillary network could also serve as a trigger to release cells from adjacent tissue by the controlled release of growth factors and suitable lead structures. In this way, tissue size may increase piece by piece. Since the capillary network is in direct contact with the embedded tissue, defined areas may be covered with proteins of the ECM, for example, to allow oriented migration of cells (guiding) and the formation of tissue structure (differentiation). With these very different materials, three-dimensional spaces can be built into tissue containers that contain information regarding biological sequences, such as cell adhesion, cell migration, cell division and interphase. Therefore, it is possible to define or shift borders between the growing tissue and the interstitial space, and to promote or inhibit the formation of surface structures of the growing tissue.

Naturally and chemically generated ECM proteins can be used as an artificial interstitium, e.g. collagen preparations of any species from the skin, bone, cartilage, horn and hoof, and from tissue preparations such as swim bladders of fish, rooster combs and trachea, for instance. Special collagens like reticulin can be prepared from individual organs. This provides the option to leave, remove or add other matrix proteins essential for tissue development, such as proteoglycans, fibronectins, vitronectins and laminins apart from the different collagen forms.

7.3
Smart Matrices

Interstitial matrices for the new culture system may also consist of recombinant matrix proteins. Any kind of fibrillar proteins with collagenous or non-collagenous properties of ECM proteins can be produced in analogy to known matrix proteins. These components can subsequently be copolymerized to form three-dimensional networks with various mesh sizes. It is possible to optimize these kinds of constructs with respect to their functional properties, unlike with naturally occurring collagens. Since recombinant collagens and other matrix proteins are constructed sequentially by single amino acids, additional information motifs may be introduced during the amino acid biosynthesis. These motifs are particularly important for cell adhesion, and they interact with cell adhesion proteins like integrins, cadherins, immunoglobulins and selectins. Such peptide sequences are formed by the RGD sequence which is found in vitronectin, fibronectin and collagen, for example. Tyr-Ile-Gly-Ser-Arg (YIGSR sequence) is found in β1-laminin. Arg-Glu-Asp-Val (REDV sequence) is part of fibronectin. With these kind of motifs, information regarding cell adhesion, cell migration, cell division and differentiation may be individually integrated, and therefore serve for the focused guidance of cells. If such a smart matrix is placed on top of a growing tissue, various cell properties may be influenced at the growth boundary. In this way, areas are defined where cells migrate into the matrix or where their migration is particularly inhibited.

Another method of forming an interstitial space could be to use organic polymers equipped with functional groups for cell interaction. RGD sequences can be attached to polyethylene therephtalate (PET), polytetrafluorethylene (PTFE), polyvinylalcohol (PVA), polyacrylamide and polyurethane. Surface modifications based on amino acid sequences may be used with hydrophilic polymers like polyvinyl pyrollidone (PVP), polyethylene glycol (PEG), polyethyleneoxide (PEO) and polyhydroxyethylene methacrylate (HEMA).

7.4
Optimal Housing for the Perfusion System

Apart from an artificial interstitium, suitable perfusion containers are required for optimal culture conditions. A particular prerequisite for the generation of an optimal construction is the ability to simulate a large variety of physiological conditions. Due to the physical separation of the artificial interstitium and the space for the growing tissue, large numbers of variations in the culture container are possible with respect to shape and size (Figs 7.1 and 7.2).

The culture containers constructed in our lab feature a basal part and a lid (Fig. 7.2). A perfusion chamber is constructed in such a way that a cell carrier or a piece of tissue is first fixed to the basal part. Subsequently, the lid whose inside is filled with artificial interstitium is attached. The basal part and lid are separated by a gasket. Either the lid or basal part has an inlet for culture media, while the opposite side of the chamber has

Fig. 7.2: A culture container consists of a basal part and a cover. The cell support is attached to the basal plate of the perfusion chamber. A cover is then applied whose interior is filled with an artificial interstitium.

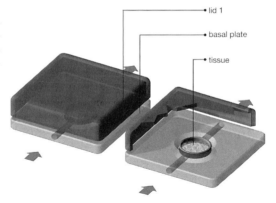

lid 1

basal plate

tissue

the outlet for the media. After assembling the lid and basal part, the chamber is tightened by applying pressure with the help of a spring-lock. The artificial interstitium is laid closely against the tissue inside the chamber. The advantage of this innovation is that the chambers are minimized with respect to height, the void volume is reduced and at the same time the media exchange is optimized (Fig. 7.2).

A gradient chamber can be formed from two identical lid parts and a modified basal part. During the assembly, a suitable tissue carrier is placed into a drill hole of the basal part to serve as a functional gasket between the luminal and the basal chamber compartments (Fig. 7.2). Another advantage of this construction is that, in principle, all common tissue carriers and filter insets can be integrated. For this, a lid is attached above and underneath the basal part. Upper and lower compartments can be filled with the same or different media. In this way, gradients may be applied to tissue in the same way as they occur under natural conditions in the organism.

Connective, muscle and neuronal tissue can be supplied with individually balanced media. Epithelia may be cultured with different media for the luminal and the basolateral side in a physiological gradient similar to natural conditions. It is possible to expose the tissue to very different liquids or gases, as is typical for epithelia in an organism.

7.5
Supply of the Maturing Tissue with Medium

Maturing tissues must be supplied with fresh medium in a continuous fashion, or in intervals, for optimal differentiation (Fig. 7.3). Preferentially, a peristaltic pump with individually selectable pump rates of a few milliliters per hour is used. The culture medium is normally transported via a tube system, which consists of a variety of different materials. However, in the tubing system, media which is maximally loaded with oxygen has to be transported for the optimal supply of the culture. Air bubbles that form preferentially at material contacts and which do not dissolve within foreseeable time intervals are problematic. When aspirating the medium, small gas bubbles

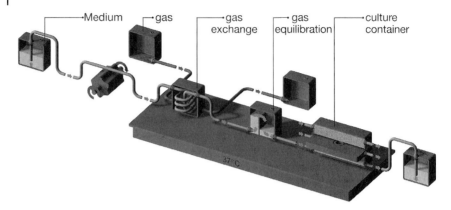

Fig. 7.3: Tissue factory – modular structure of components for the improved production of tissue constructs.

are formed that are not recognizable in the liquid column with the naked eye and which increase substantially during transport of the media. Like an embolus, these bubbles will finally obstruct the progression of the media. The air bubbles will then enter the downstream culture container at irregular intervals. Areas containing air bubbles will not be supplied homogenously with nutrients.

For the reasons described, in our opinion, the gas-saturated culture medium has to be supplied by a tubing system that consists of uniform material and has as little volume as possible. For this reason, a novel plug for the supply bottles was developed that has several passages and which prevents contamination of the interior of the bottle. For minimization of gas bubble formation, one of the special passages is equipped with a tube that reaches the basal part of the container which is used to aspirate the media and also serves as a pump vessel. A second tube enters by another passage and serves for filling or venting the container. It is also possible to add a sterile filter so that sterile gases enter when the container is emptied. For the first time, only a single material without any seam is now used from the culture media supply bottle, up to the container and the collecting reservoir. This prevents the formation and accumulation of air bubbles.

A gas mixture with positive pressure is frequently applied to a storage vessel of the culture medium for oxygenation and for the stabilization of the pH. The disadvantage of this method, however, is that gases in the form of different size bubbles accumulate in the culture medium. In the case where this gas-rich culture medium is transported to a culture flask via a tube system, gas bubbles emerge during the transport process, collect at arbitrary places in the system, replace the liquid, and result in pressure variations and disturbance of the homogenous flow through the culture flask. When using this method, contamination is frequently observed, introduced to the culture media by an impure gas supply. The technical dilemma in a perfusion culture is the discrepancy of optimal oxygen supply to the culture media and the need to prevent the bubbling out of gases within the construction, since areas with air bubbles are not homogenously supplied with media. Therefore, a module that provides a simple way for any kind of gas to enter the sterile culture media and that prevents bubble formation of the media in the culture flask is necessary.

In our method of gas enrichment (Fig. 7.3), culture medium is pumped slowly through a long tube with a small inner diameter. The wall of the tube is gas-permeant and may, for instance, be made of silicon. The material guarantees optimal diffusion of gases between the culture media inside the tube and the surrounding atmosphere. By means of the wall diameter, the inner diameter and the length of the tube, oxygen or other gases, such as carbogen, the media can be enriched with gas in a simple and reliable fashion by diffusion.

The gas-permeable tube, with a defined inner diameter, wall strength and length, is wound up in a holder. The spiral tube is then put into a lockable container with an inlet and outlet for gas. Any kind of gas mixture can be applied to the spiral tube at various intensities. If culture media is pumped through the spiral tube, the gas accumulates in the culture media by diffusion over the tube wall. Thus, any kind of gas may be applied to the culture media under absolutely sterile conditions. At the same time, the pH of the culture media may be regulated, bubble-free, without complicated injection equipment for any period of time.

The release of gas bubbles from the culture media can be minimized by a gas header tank.

In addition, the gas-saturated culture media is pumped to the special container on the bottom side. The medium is lifted up in a channel inside the container. After a short while the channel ends. The media is then able to spread and equilibrate in a gas-filled chamber.

Gas bubbles will be released in this area of the medium. The liquid collects again in a funnel-shaped area and is transferred to the culture container via a tube. The medium is now perfectly enriched with gas, but is free of bubbles.

The header tank is vented at its upper side. This exhaust can be attached to a second, parallel expansion tank. In this way, two channels of a gradient container are interconnected by a gas bridge, resulting in an unrestricted passage of bubble-free media in between the luminal and basolateral compartments.

Most tissues have to be generated at a constant temperature, preferentially at 37 °C. For this purpose, the gas-enriching module, gas-balancing module and tissue container can be placed in an incubator. However, this process has disadvantages, since the individual parts are not easily accessible from all sides. For this reason, it is more favorable to hold the modules on a regulatable heating plate, placed on a table where it is easily accessible from all sides and provided with a cover to prevent temperature loss. It is particularly favorable if an electronically regulated heating device is built directly into the modules. This is easily realized by inserting or attaching a heatable and adjustable thermal foil to the wall of a module. To avoid electric shock, the heating device should equipped with a low-voltage power supply and a safety switch. The heating module is not located in direct contact with the culture media or the inserted tissue, in order to avoid contact to inappropriate materials.

The individual tissue culture components are assembled into a workline (Fig. 7.3). From a storage vessel, the culture medium is transported at around 1 ml/h via a tube and a peristaltic pump into the gas exchange module. Emerging gas bubbles can be eliminated in the attached gas-exchange module before the medium reaches the tissue container. Spent culture medium is collected in a waste container and is not recirculated. The gas-exchange module, gas-balancing module and culture container are placed on a heating plate, which keeps the surrounding environmental temperature constant. The heating plate and the assembled modules are covered by a protective acrylic cover during the course of the culture. As soon as the buffer of the media is adjusted to the type of gas in the gas exchanger, the system may be operated for long periods of time under very reproducible conditions, without contamination (Fig. 7.3). Work-lines with various configurations may be very easily operated in parallel with one peristaltic pump (Fig. 7.3).

7.6
Synopsis

A concept for a modular system is introduced in which different kinds of tissue mature under physiological conditions and are maintained over long time periods in a differentiated form. Newly developed culture containers with artificial interstitia, special transport technology for the culture media as well as novel gas-exchange and -balancing modules for an oxygen-rich, bubble-free supply of the constructs are introduced.

8
Ensuring Tissue Quality

Tissue construction progresses through various subsequent stages from start to finish. The technical procedures involved – such as pipetting, media preparation, tissue incubation and record keeping – underlie clearly defined quality norms and can be reproduced. In contrast, where the assessment of the degree of maturation and functionality of the tissue construct itself is concerned, quality control seems to be insufficient or even totally lacking.

From a surgeon's point of view, it does not really matter whether the generated implant is mature or not, as long as the healing process is smooth and the lost function is replaced. One can sympathize with this view, but the other side of the coin is that clinical experience is very limited with regard to the implantation of tissue constructs grown in cultures. It will take many years, if not decades, until reliable statements can be made on the cell biological suitability of implants for patients. When metal or polymer implants were first developed it also took at least a decade of continuing optimization to reach our current level of knowledge and quality standards. The same applies to artificially generated tissue. The key for future success lies in our ability to regulate tissue-specific differentiation in the developing constructs.

[Search criteria: quality control tissue engineering]

8.1
Norms and Cell Biology

Formulating guidelines for the quality management of tissue constructs seems to be a particularly difficult task. It is not simply a matter of defining the work environment with its specific tools and procedures, as in standard operation procedures, but of recognizing the apparent as well as the latent developing potential of the tissue constructs that are going to be implanted into patients. The quality of the construct itself, including cell biological differentiation, does not feature in any quality norms and regulations known to us. This makes it all the more urgent to investigate if data that have been mostly obtained from mouse stem cells can be equally applied to the human race. If no major developmental discrepancies show up, we will soon be able to make therapeutic use of them. If, however, there are fundamental differ-

Tissue Engineering. Essentials for Daily Laboratory Work W. W. Minuth, R. Strehl, K. Schumacher
Copyright © 2005 WILEY-VCH Verlag GmbH & Co. KGaA, Weinheim
ISBN: 3-527-31186-6

ences, this knowledge – in connection with a growing sense of critical awareness – can be put to good use to find a new viable strategy. It is becoming clear that it is not the technological aspect, but the cell biological potential in the cultured tissue cells that will set the pace of further progress and tell us where to go from here.

New strategies must be found for the culturing of tissue constructs, since many of the methods developed so far have proved to be of little use. Furthermore, what is needed is more reliable knowledge about tissue development within an organism. This might enable us to transfer nature's proven development strategies to *in vitro* tissue cultures and thus make significant improvements to the quality of constructs. This is an area where science has some catching up to do. While recently, many surprising discoveries have been made regarding the functional control of individual cells, comparatively little is known about the molecular processes involved in tissue development. Such knowledge, however, is indispensable if progress is to be made in the professional production of artificial tissue and we should concentrate our research efforts on this particular area. In terms of research policy, the insight must prevail that the problems we have encountered regarding the regulation of cell differentiation processes can only be resolved if adequate research programs are put in place.

Many people feel they are in a position to discuss or even pass judgment on matters regarding stem cell differentiation. A large part of the media as well as certain factions in the scientific community are telling the public that there are no risks involved in growing any kind of tissue from embryonic stem cells. At the present moment, such a view is over-optimistic rather than objective. The fact is that, so far, stem cells kept in culture dishes have only been able to produce monolayers of precursors of functional neurons, connective tissue, muscle or epithelial cells. A monolayer of precursor cells is a far cry from functional tissue and, up to now, there has not been a single case of such cells developing into functional tissue. This remains a task for researchers in the coming decade.

The difficulties and challenges that still remain concern the development of truly functional three-dimensional tissue from cultured cells with the help of microreactors. The resulting construct should be graspable with tweezers and implantable. It should also provide or find a perfect connection to the vascular system. It is not self-evident that constructs will develop into functional tissue, as experts know only too well.

[Search criteria: ISO quality cell culture]

8.2
Evaluating Complexity

When looking at the production of tissue constructs, all factors that may affect cellular differentiation and possible variation in the biological make-up of the construct must be taken into account. As was shown in the schematic example of the hematopoietic system, embryonic cells progress through several intermediate stages before developing into functional, but isolated blood cells (Fig. 8.1). We know that the steps in this process are irreversible. An erythrocyte cannot revert into a pro-erythroblast (Fig. 8.1).

Hematopoiesis

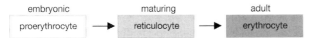

Fig. 8.1: Terminal differentiation of an erythrocyte. During *in vivo* erythropoiesis, functional cells develop from embryonic precursor stages. This process is irreversible.

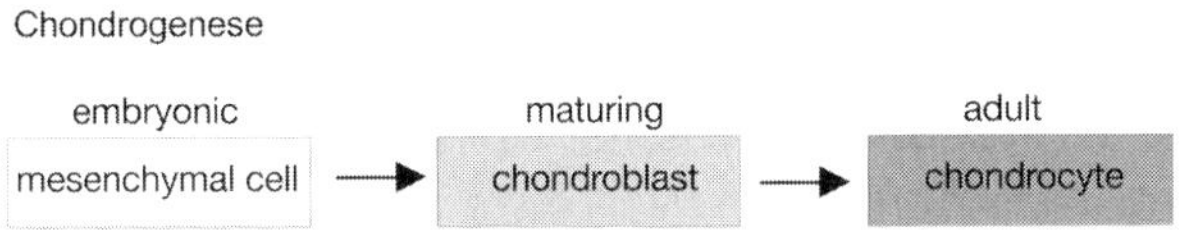

Chondrogenese

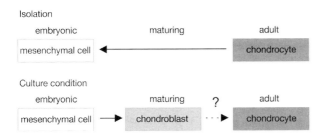

Fig. 8.2: Terminal differentiation in a chondrocyte. During natural development, a mesenchymal cell turns into a chondroblasts and then into a chondrocyte. This is an irreversible process.

The rule of one-directional development applies to most, but not all, tissue cells in the body. From a mesenchymal cell, for example, arises a chondroblast which turns into a chondrocyte (Fig. 8.2).

In contrast to chondrogenesis (Fig. 8.2), the artificial development of cartilage tissue is a very different process. The shape of a chondrocyte found in a cartilage capsule is round. After it has been isolated, it turns into a flat cell type resembling a mesenchymal cell or a fibroblast, (Fig. 8.3), but this does not make it a native mesenchymal cell.

Using a scaffold, attempts have been made to revert the fibroblast-like cell into a chondroblast and, if possible, into a chondrocyte able to produce resilient cartilage ground substance. Many studies have shown that not only tissue-characteristic chondrocytes are found, but many intermediate stages ranging from fibroblasts and chondroblast to the adult cell type.

Isolation

Culture condition

Fig. 8.3: Re-embryonalization of a chondrocyte into a mesenchymal cell in tissue engineering. When a chondrocyte is isolated from adult tissue and grown under *in vitro* conditions, it turns into a cell that resembles a mesenchymal cell, but is not identical to such a cell. What is atypical in this cell is that it is grown under culture conditions in order to produce as many chondroblasts as possible. A cartilage cell within cartilage tissue would not have done that. It is not clear yet if such a cell could produce a functional chondrocyte.

In contrast to the natural development of cartilage, the developmental stages in the generation of artificial cartilage tissue under *in vitro* conditions are reversible. A chondrocyte can be isolated from cartilage through the degradation of ground substance and then grown in a culture (Fig. 8.3). In contrast to a chondrocyte embedded in ground substance, such a cell can now be propagated and used to build an artificial cartilage construct. In the process, the isolated cell has completely changed its shape. It is no longer round like a typical chondrocyte, but flat, looking like a fibroblast. For reasons yet unknown, the synthesis of cartilage-typical collagen type II has been switched to the production of atypical collagen type I. As a consequence, the secreted collagen monomers can no longer connect to an ECM that is able to withstand physical strain. The intercellular substance becomes uncharacteristically soft, which limits its suitability for joint surface replacement.

It is also true that cartilage does not always equal cartilage. From a histological and functional view, three types of cartilage can be distinguished, they are found in specific locations and nowhere in the literature have the procedures for the production of hyaline, elastic or fibrocartilage been described. An outer ear, for example, must have elastic properties and cannot be replaced solely by inelastic hyaline cartilage. The type of elastic cartilage that would be required has not yet been generated *in vitro*. There are also no data available that would explain why the number of chondrocytes inhabiting a cartilage lacuna varies so greatly. No research results have been available to shed light on the formation of cartilage capsules that form the border between chondrocytes and ECM. Finally, often, no suitable marker can be found to detect differences in differentiation, and immature cells in the tissue cannot be distinguished from maturing and matured cells. Such gaps in our knowledge do not only apply to the generation of cartilage, but also to other tissues of the body. It is very strange indeed that although we have known for nearly a century what the various developmental stages of organs and muscles are, the developmental physiology of the tissues involved – the maturation of tissue and the development of functional structures – is largely unknown to this day.

[Search criteria: differentiation functional *in vitro*]

8.3
Expression Behavior

Trials are already under way to grow a variety of tissues under culture conditions by placing cells on a scaffold. The resulting constructs are not perfect and more often than not subject to dedifferentiation. Experiments also show that not every scaffold is equally suitable for cells to settle on and some may well interfere with the development of optimal tissue structures. It is therefore important to keep a critical eye on the growth behavior of cells and emerging differentiation. Above all, it must be investigated whether atypical structures develop alongside typical structures – in other words, adequate quality monitoring of cultured cells and tissues must be implemented. The aim should be to generate a construct that is functionally equivalent

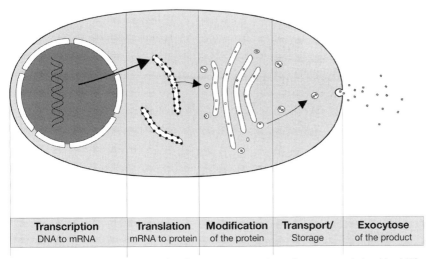

Transcription	**Translation**	**Modification**	**Transport/**	**Exocytose**
DNA to mRNA	mRNA to protein	of the protein	Storage	of the product

Fig. 8.4: The proteins of a cell are produced on a transcriptional as well as on a translational level. When artificial tissues are generated under less than optimal conditions, the synthesis chain can be interrupted at various levels.

to the corresponding structures in the living organism. Only such comparisons will enable us to decide if a tissue culture can be classified as a success or failure.

Each cell in a tissue carries information in its DNA that is transcribed into mRNA during cellular differentiation and then transported into the cytosol. The information goes to the ER and the ribosomes where the individual nucleic acids are synthesized into proteins. These are then used for internal cell tasks or secreted. The information transfer from DNA to mRNA is called transcription, while the transformation of mRNA into proteins is known as translation (Fig. 8.4).

The information contained in mRNA is not automatically transformed into functional protein. Being an intermediate link in the protein production chain, mRNA can only give unreliable information about the functional protein that will be available at the end of the process. The end product, i.e. protein, is the result of a complex chain – involving, for example, the Golgi apparatus which, by attaching sugar residues to the molecule, conveys the functionality of a glycoprotein to the protein. The possibilities of shaping the functional properties of a protein through folding, glycosylation or phosphorylation are manifold.

Especially in cell and tissue cultures, normal processing of proteins is not guaranteed, and successful transcription does not necessarily imply immediate and complete translation. It is amazing how much importance is given to mRNA findings – especially in cell or tissue cultures – and how little is known about the translational level – the regulation of protein biosynthesis. However, rapid change can be expected, due to the progress of current research into proteomes.

When testing the quality of a cell or tissue – also called profiling – the general question arises if detection of the synthesized products should be carried out at a transcriptional or translational level. Of course, it is always best to establish a profile

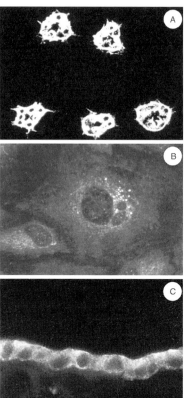

Fig. 8.5: Loss of antigen expression, caused by suboptimal culture conditions. Immunohistochemical labeling of collecting duct epithelium in a kidney (A), of a monolayer of collecting duct cells cultured in a cover slip (B) and of collecting duct epithelium cultured on a kidney-specific matrix (C). The monoclonal antibody used as label clearly identifies the cells in the collecting duct system of the kidney. Collecting duct cells growing as a monolayer on a cover slip are marked as dots, which is atypical (B). Although the protein in question has been produced, it cannot be transported to the plasma membrane. All cells of generated collecting duct epithelium grown on a kidney-specific substrate show the labels very clearly (C).

at both levels – if suitable methods and markers are available. Where known proteins are to be detected, this is a quite straightforward procedure. PCR is a biochemical tool that detects at a transcriptional level whether a cell produces the corresponding mRNA. In order to clarify at a morphological level whether the mRNA has been produced by the cells under investigation and not perhaps by neighboring cells, *in situ* hybridization is carried out on tissue sections that then undergo light and electron microscopy. Finally, using electrophoresis, protein fractions can be separated from the cell. The presence of a protein as a marked band can be shown in Western blotting with the help of its antibody. All these methods provide sensitive and very reliable tools to show if specific proteins are synthesized under the chosen culture conditions or not.

When tissue constructs are grown, they should be compared with the corresponding organic tissue in as much depth as possible. Not only adult tissue, but also embryonic and especially semi-matured tissue structures should be included in order to obtain as much information as possible about the ongoing developmental processes in the culture. Through such comparisons, it could be shown that cells that had been isolated from a kidney and were growing as a monolayer on a cover slip contained the mRNA that was needed for the synthesis of certain proteins. However, although they were able to synthesize them, they could not integrate the synthesized protein into the plasma membrane under the given culture conditions. The reason was found to be a change in the shape of the cells which were cultured on an unsuitable substrate (Fig. 8.5). Cells that were originally isoprismatic have turned into atypically flat cells (Fig. 8.5B). PCR and Western blot would show positive, but incomplete, results for such a protein. The protein has been synthesized, but – due to the wrong shape of the cell – it seems to be in the wrong location, as it cannot be detected in the plasma membrane. Such important findings can only be obtained if adequate morphological, molecular biological and immunohistochemical methods can be combined as required.

Culture conditions should generally be chosen in such a way that the cells in a construct are not only able to synthesize a protein, but also to process it in a tissue-specific way that allows its specific functions to develop (Fig. 8.5C). For this reason, we recommend that when a culture experiment is set up, Western blotting should be carried out first in order to detect the presence of a specific protein at the translational level. This should be followed by immunohistochemical analysis to establish whether the protein is found at the right location. A positive result bodes well for the success of the culture.

[Search criteria: cell features culture detection]

8.4
Suitability of a Scaffold

When cells are cultured on a scaffold, it is particularly important to find out whether the cells are evenly distributed over the whole surface – which would make them suitable epithelial cells – or if they prefer to settle in a three-dimensional space – which would make them suitable connective tissue or neuronal cells. Further development of quality tissue depends largely on the functional interaction between cells and the biomaterial used, and on the typical distribution of cells. The reaction during primary and secondary contact shows if epithelial cells adhere perfectly to the implant and will thus be able to withstand rheological stress – as caused by the bloodstream (Fig. 8.6).

For connective tissue cells, such as cartilage cells, primary contact with the artificial matrix should result in a homogenous distribution of chondrocytes – which would enable them to build load-bearing intercellular substance (Fig. 8.7). If cells only settle on a few places on the scaffold during secondary contact, these would be the only places where typical intercellular substance could develop. This would result in an inhomogeneous matrix that could not withstand physical strain and be practically useless as an implant.

Several test series must be carried out to find the best-suited material for a specific tissue – with cell lines first, then with primary cultures. The cells are grown on a variety of scaffolds held in tissue carriers. After certain defined culture periods, the tissue carriers are fixed and stained with fluorescent nuclear dye (Fig. 8.8). A microscope in epifluorescence mode will show the distribution of the cells on the scaffold – independent of the transparency of the scaffold itself. All modern fluorescence mi-

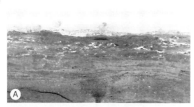

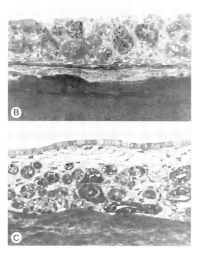

Fig. 8.9: The scaffold used has an effect on the development of renal epithelial and connective tissue. (A) Scaffold A shows that neither the epithelial nor the connective tissue are clearly developed. (B) Scaffold B clearly shows that connective tissue has developed, but not epithelial tissue. (C) Scaffold C gives distinct support to the growth of connective tissue as well as luminally limiting epithelial tissue.

the development of epithelium, but also the formation of the underlying connective tissue (Fig. 8.9A). While scaffold B inhibits the formation of epithelium, it supports the development of lamina propria (Fig. 8.9B). Only in the naturally occurring scaffold C can both the epithelium and its lamina propria develop. (Fig. 8.9C).

[Search criteria: scaffold cell distribution fluorescence]

8.5
Hidden Heterogeneity

The challenge of producing artificial tissue lies in the development of specific properties and the avoidance of dedifferentiation. All cellular and extracellular characteristics of a specific tissue should be permanently present, if at all possible. Clear guidelines must be put in place that define the criteria by which the structures of cultured tissue are compared to those found in an organism. These must include thorough cell biological and pathological evaluation. Such profiling can detect which properties are present, which are missing and what atypical properties have emerged.

Frozen sections provide a fast and effective way of detecting specific tissue properties at a morphological level. The tissue that has either been taken from an organ or artificially created is put on a surface and covered with Tissue Tec as quickly as possible and then frozen in a tissue holder under extensive use of CO_2. The tissue holder is then clamped into a freezer microtome, and very thin (around 5–10 μm) sections are cut and mounted on a glass slide. The sections are then dyed with 1% toluidine blue solution, dehydrated with alcohol, cleared with xylol and finally put under a cover slip.

A microscope will quickly identify living cells and developed tissue, while answering further important questions at the same time, e.g. if an epithelium contains isoprismatic or atypically flat cells. Likewise, it can clarify whether cells are growing in a monolayer or in multilayers and if they grow closely together or keep a discrete distance from each other. Changes in the position of nuclei, the shell shape or the structure of the layers give additional first clues regarding properties such as polarization, transport capacity or resilience of tissues.

Using indirect immunofluorescence, the presence and distribution of certain proteins in a tissue can be proven beyond any doubt. A cryostate section is prepared and fixed in 100% iced ethanol for 10 min, then washed in PBS for 2 × 5 min and incubated in blocker solution for 30 min (PBS + 1% BSA, bovine albumin + 10% horse serum, HS), in order to saturate non-specific binding sites. The blocker solution is siphoned off and the preparation incubated in primary antibody solution for 90 min. The solution contains the antibody to the protein to be detected in the tissue section. Then the preparation is washed again in PBS + 1% BSA for 2 × 5 min and incubated in FITC secondary antibody solution for 45 min. The secondary antibody, which has been stained with a fluorochrome, binds to the antigen–primary antibody complex, thus making it visible. From now on, the incubation must be protected from light. Finally, the preparation is washed again in PBS + 1% BSA for 3 × 5 min before it can be transferred to a slide and embedded. Immunofluorescence can be analyzed using an epifluorescence microscope at a stimulation wavelength, e.g. 495 nm.

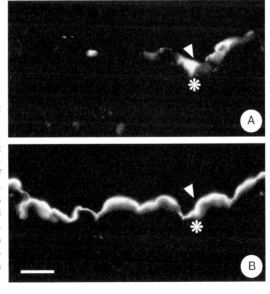

Fig. 8.10: Expression of individual proteins in tissue. Renal collecting duct epithelium has been generated in IMDM (A) and IMDM + NaCl (B). In both cases, fully structured collecting duct epithelium with polar differentiation developed. On the grounds of purely morphological criteria, no difference can be found between the two epithelia. Only immunohistochemical labeling with mAb 703 makes the differences apparent. While the culture in IMDM gives rise to only a few antibody-binding cells, all cells grown in IMDM + NaCl are carrying the label.

A light microscope is often unable to detect the difference between tissue samples grown under different conditions, so it is worth investigating differences in the expression profiles of proteins immunohistochemically. This can be done very effectively using frozen sections.

For example, renal collecting duct epithelium can be generated in IMDM or in IMDM plus 12 mmol/l NaCl (Fig. 8.10). In both cases, contiguous, collecting duct epithelium with polar differentiation develops and no difference between the two epithelia can be discovered under light microscopy. Only immunohistochemical labeling with mAb 703 shows up clear differences (Fig. 8.10). While only very few antibody-binding cells have developed in the culture using IMDM only, all epithelial cells in the culture using IMDM + NaCl are carrying the label. These findings can now be compared with the expression profile of collecting duct epithelium from a kidney. Only on the basis of these results can a judgment be made whether the status of the cultured epithelium is typical or the protein has been hypo- or hyperexpressed.

[Search criteria: cell culture heterogeneity expression]

8.6
Investigating Cellular Ultrastructures

When profiling tissue constructs, light microscopy techniques are often insufficient. Transmission electron microscopy (TEM), in contrast, can achieve a higher optical resolution of cellular structures, and thus give an insight into the subcellular distribution of organelles, the constituents of the cell membrane and the basal membrane, as well as into surface differentiation and cell contacts. However, the preparation of an electron microscopic sample is far more labor and time intensive, and thus more

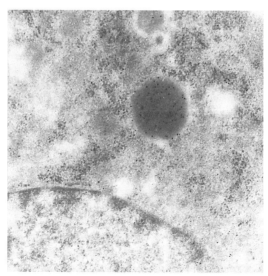

Fig. 8.11: Electron microscopic/immunohistochemical detection of a renin-containing granule in cultured kidney tissue. The gold marking shows up as a small black grain found exclusively within the granule.

expensive, than the preparation of a section for light microscopy. What gives electron microscopy the edge is its unsurpassed ability to identify intracellular structures. Vital questions regarding the onset of surface differentiation and cell polarization can only be answered through electron microscopy. This includes the visual representation of product excretion and directional transport functions. Electron microscopy has the ability to deliver a precise topological image of cell organelles, including the Golgi complex.

With the help of electron microscopy and, above all, immunohistochemical methods, not only cellular polarization, but also the orientation and content of organelles within the cells can be clearly recognized (Fig. 8.11). The secondary antibody used in the procedure is marked with gold instead of fluorochrome, which will makes them appear as electron-dense dark dots on the TEM screen. Detailed analysis can now be carried out to find out, for example, if the trans-Golgi reticulum has an intact three-dimensional structure, reaching the surface of the epithelium – which would allow proteins to be processed along that route.

While it may well be that under a light microscope, the expression of tight junction proteins such as ZO1 or occludins has been immunohistochemically detected through marking with their antibodies, it is still possible that no functional, i.e. sufficiently tight, junctions develop in the construct. The reason for this may lie in an insufficient

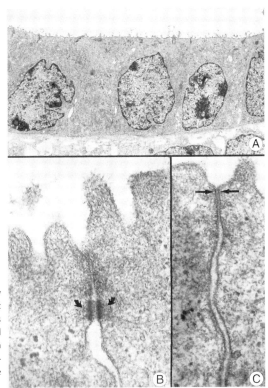

Fig. 8.12: Electron microscopic view of generated renal collecting duct epithelium. (A) The epithelium has found a basal membrane on its basal side. (B and C) At the border between the apical and the lateral plasma membranes, functional tight junctions are clearly visible.

number of anastomosing single strands. Normally, five or six such strands are found (Fig. 8.12), but if only three to five strands have developed, a tight junction will be no longer physiologically tight. The typical structure of functional tight junctions in an epithelium can only be properly evaluated under an electron microscope with the help of a freeze-fracture replica.

[Search criteria: transmission electron microscope ultrastructure tissue]

8.7
Functional Transfer

8.7.1
ECM and Anchoring

Tissue cells synthesize their ECM or basal membranes either by themselves or in close cooperation with neighboring tissues, which results in the formation of specific compartments. These keep the cells and sometimes also tissue at a certain distance from each other, bunch them together in groups or separate them. The consistency of ECM and its properties are tissue specific. While the basic ECM components such as fibronectin, collagen and proteoglycans are always present, their actual composition may vary, and there are more than 20 different types of collagen and collagen-like molecules that – through polymerization with fibronectin and proteoglycans – allow for an infinite variety of three-dimensional cross-linking. The amino acid sequence of these molecules also contains information regarding cell adhesion and motility. The presence of all these proteins can be detected with cell biological methods and antibodies. Through consecutive immunoincubations performed on electron microscopic sections or through computer-aided reconstruction, the involvement of individual proteins and their cross-links can be shown. The picture that seems to emerge indicates that the ECM of each organ and its specific tissues has its very own characteristics.

While some tissue cells have a rather loose contact with their ECM, others are in very close contact (Fig. 8.13). Where the plasma membrane of a cell makes contact with the ECM – this could be just a focal point or a larger surface area – cell-anchoring proteins are established. These are integral membrane proteins to which specific amino acid sequences of the ECM can bind. They are known as integrins and have a heterodimer molecular structure, consisting of an α and a β subunit. There are at least eight different types of α and β units, which can be combined in a variety of ways to bind to a wide range of structures in the ECM. This explains why a difference in the composition of α and β subunits enables tissue cells to bind to specific components in the ECM. The configuration of α and β subunits within a tissue is not constant, but varies depending on the degree of maturation.

In endothelial cells, integrins are responsible for the adhesion of the cells to the vascular walls, thus ensuring that the cells are firmly attached to the matrix and not washed away by the bloodstream. When, for a vascular prosthesis, biomaterial is inoculated with endothelial cells, it is possible that typical integrins are expressed

Fig. 8.13: Tissue-specific anchoring reaction between ECM and cells. The diagram shows the anchoring of epithelial cells to the basal membrane (A) and of a connective tissue cell to ECM (B). In both cases, very specific integrins are expressed.

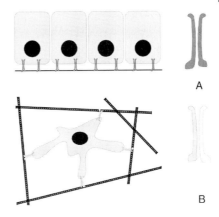

A

B

insufficiently or not at all, or atypical integrin dimers might emerge. If this happens, the endothelial cells cannot adhere to the chosen biomaterial; thus, the surface structure of the biomaterial must be optimized until the endothelial cells can develop their specific cellular anchors. The same applies to bone cells (osteoblasts). They will only begin producing a collagen type I network, followed by a calcification process, once they have been well anchored via their integrins.

This last example shows that the anchoring process does not simply result in a physical contact between cells and the ECM, but may also establish a functional link to the cellular metabolism. Such processes are regulated by kinases of the ERK group and MAP kinases. This functional cascade, which depends on the ECM, regulates cell adhesion, cell division and the length of the functional interphase. As it is very difficult *in vitro* to regulate the inner cell processes externally, all cell functions depend on the interaction between cell and biomaterial alone. Suitable biomaterial and tissue-specific integrin expression are indispensable for the production of functional tissue in an experimental culture.

[Search criteria: extracellular matrix integrin signal transduction kinases]

8.7.2
Development of Cell–Cell Contacts

When generating tissue such as epithelia, their polarization behavior should be evaluated. This involves finding out whether the right proteins – those that would allow a natural functional barrier to develop – have been integrated into the apical or basolateral plasma membrane. Epithelial cells are characterized by the development of tight junctions as a result of lateral cell–cell contacts. By sealing off intercellular spaces, epithelia prevent the passage of molecules between the cells, thus ensuring that only transcellular transport of molecules is possible. Thus, it depends on channel structures and transport proteins in the luminal or basolateral plasma membrane which molecules can enter the cells and which molecules are refused entry at the epithelial barrier.

The 24 members of the claudin family are important structural and functional components of tight junctions – known as occludins and the junctional adhesion molecule (JAM). In analogy to integrins, corresponding variable claudin pairs form at tight junctions. Additionally, there are proteins associated with tight junctions which can be easily detected using immunohistochemical methods or with the help of an antibody in Western blotting. However, the detection of occludins alone does not give any information about their functional sealing properties, which need to be examined physiologically and morphologically.

Apart from tight junctions, gap junctions are also essential mediators of functional cell–cell contacts in epithelial and non-epithelial tissue structures. They enable the exchange of substances from one cell to another. Gap junctions consist of two corresponding channel structures (connexons) that go through the plasma membrane. These, in turn, contain six similar connexins. We know now a wide range of different connexins that can shape the exchange of substances and information between cells in many different ways. Again, antibodies to the amino acid sequences of individual gap junction proteins (connexins) can be used to verify their tissue-specific expression. The connection between heart muscle cells in the intercalated disks, which regulate the exchange of small molecular substances, and the transmission of electrophysiological stimulation within the heart muscle, is a striking example. This is where the contraction of all connected cells is synchronized. For obvious reasons, cardiomyocytes grown *in vitro* must develop a sufficient number of gap junctions in order to be functionally connected to neighboring cells after implantation. Cells that are implanted into a contractile tissue must be grown on an elastically deformable ECM. They must also have information motifs on their surface or in the closely pericellular matrix that support the growth of capillaries into the construct at the highest possible speed to ensure the tissue is integrated into the vascular system.

Gap junctions come in a wide variety of structures and are found not only in epithelia, but also in embryonic, maturing and adult connecting tissue. Thus, even the functions of physically separated cells can be coupled through long extensions, and information can be exchanged. As in epithelial structures, gap junctions are found on a regular base, connecting the cytoplasm of neighboring cells. Thus, synchronized functions can be optimized, which is crucial, for example, in the building of matrix for a long bone. The process involves the calcification of large areas that are vitalized by a communicating network of osteocytes. In neural tissue, another, very different, communication strategy is developed. Information is passed along extremely long dendrites and axons. Incoming information must be processed or bundled, which is done at the synapses between neurons and/or an effector organ.

[Search criteria: cell contact gap junctions connexin review]

8.7.3
Cytoskeleton

All cells have a cytoskeleton that consists of actin filaments, intermediary filaments and microtubules. The structure and location of these structures vary, depending on the type of tissue cell. The cytoskeleton is a three-dimensional structure within the cell, thus forming its endoskeleton. Its components form a non-static, elastically deformable structure which retains the shape of the cell and has a supportive function when it comes to positioning organelles, modulating movement and forming transport routes within the cell.

The importance of the cytoskeleton in connection with the formation of transport routes becomes very clear when looking at the example of the web of microtubules in neurons. In neurons – as, for that matter, in other cells – protein synthesis is linked to the nucleus, the ER and the Golgi apparatus. These cell elements are found in the perikarya of the neurons, which are – in some cases – up to 1 m away from the distal end of the muscle-innervating axon. This makes the transport route for substances such as transmitters unusually long. Microtubules running alongside the axons are used as transport routes, with motor proteins regulating all binding and motion processes. The motor protein kinesin takes its load to the plus end of a microtubule, while the motor protein dynein takes it to the minus end. Thus, interaction between motor proteins and microtubules is involved in every transport process. What makes neural tissue functional is, among other things, the presence of the proteins mentioned above. This can be easily verified in cultured tissue, using immunohistochemical methods.

Intermediary filaments vary from tissue to tissue. Those typical for epithelial cells, for example, belong to the large group of cytokeratins (Tab. 8.1). Experiments have shown that each epithelium has its specific set of cytokeratins. In other tissues, such as muscles, there is an equivalent – desmin. In astrocytes, glial fiber acidic protein (GFAP) has been detected, while neurofilaments could be immunohistochemically identified in neural cells. Many mesenchymal tissues contain vimentin.

In cultural experiments involving epithelial cells, the detection of certain cytokines can be very helpful. For example, cytokeratin 1 is found in the collecting duct, but not in other tubular structures of a kidney. By using cytokeratin 1 antibody, it is therefore easy to verify immunohistochemically if only a single cell type is growing in a primary cell culture. It can also be used in existing cultures to find out if the production of tissue-specific cytokeratin is being maintained or if it has been replaced by an atypical product, i.e. if cellular dedifferentiation has taken place.

Cellular differentiation or dedifferentiation in growing tissues can be easily detected by immunohistochemical means with a set of antibodies. Typically, in a first step, a pan-antibody to cytokeratin is used in order to find out whether any cytokeratins at all are expressed in the cell. This antibody does not indicate if the expressed cytokeratin is tissue-specific. However, a positive reaction shows that the growing cell is an epithelial cell. Further antibodies to very specific cytokeratins are available to identify every type of maturing epithelium (Tab. 8.1).

Tab. 8.1 Presence of a wide range of cytokeratins in individual epithelia.

Type of cytokeratines	Found in
1	Epidermis, cervix
2	Epidermis, cervix
3	Cornea
4	Sebaceous glands, cervix, esophageal epithelium
5	Epidermis, sebaceous glands, sweat glands, tracheal epithelium
6	Epidermis, sweat glands
7	Sweat glands, mammal glands, kidney, urothelium
8	Sweat glands, trachea, urothelium, intestinal epithelium, hepatocytes
9	Epidermis
10	Epidermis
11	Epidermis
12	Cornea
13	Cervix, esophagus, tracheal epithelium, kidney
14	Sebaceous glands, mucous membrane of the tongue, exocrine glands
15	Exocrine glands, tracheal epithelium
16	Mucous membrane of the tongue, epidermis
17	Hair follicles, mammal glands, tracheal epithelium
18	Kidney, urothelium, intestinal epithelium, hepatocytes
19	Kidney, urothelium, intestinal epithelium, exocrine glands

Unequivocal proof of the expression of specific cytokeratins in an organism can be obtained through frozen sections of the tissue in question. If the same cytokeratins are also found in cultured cells, differentiated epithelium may develop, whereas their absence would indicate cellular dedifferentiation.

[Search criteria: cytokeratin epithelia cytoskeleton]

8.7.4
Plasma Membrane Proteins

The functionality of a tissue depends on components in the plasma membrane, such as channels, carrier and pumps. Their expression can be detected during the culturing process, using molecular biological, pharmacological and immunohistochemical methods. Furthermore, it can be deduced from their location whether they have been correctly luminally or basally integrated into the plasma membrane. As tissue properties do not automatically develop under cell culture conditions and many devia-

Fig. 8.14: Generating heterogeneous collecting duct epithelium containing light principal and dark intercalated cells in a perfusion culture.

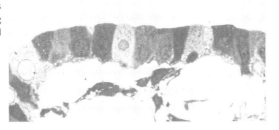

tions are possible, it is essential to carry out an exact phenotypic as well as functional characterization of the plasma membrane in the construct at a protein translation level.

We would like to demonstrate this on a generated collecting duct epithelium, the only tubular epithelium in the kidney consisting of a range of very diverse cell types (Fig. 8.14). Our main interest is to find out if the Principal Cells as well as various types of intercalated cells can develop in a culture and if characteristic membrane proteins can be detected. A major feature of Principal Cells is the epithelial Na^+ channel (ENaC) and a water channel (aquaporin 2) on their luminal side. These molecules are part of an important hormonal regulation mechanism that controls the sodium metabolism and water excretion in the kidney. On the basolateral plasma membrane of the Principal Cells, Na/K-ATPase and aquaporins 3 and 4 are found.

The α-type subordinate cells express H^+-ATPase on the luminal side. Thus, acid equivalent can be secreted to the urine as required to maintain the acidity of the body. Carbonic anhydrase type II is found in their cytoplasm, which produces H^+ ions.

By contrast, β-type intercalated cells carry luminally the anionic exchanger type I, through which urine can be made alkaline. Thus both cell types trigger the acid–base balance. Suitable markers can clarify beyond doubt if the naturally occurring functional proteins are also present in epithelia generated *in vitro*. In a best-case scenario, the expression of these structures under culture conditions would be detected on a transcriptional as well as on a translational protein level. However, this would still not tell us anything about their real functional properties. We still do not know if the transport routes are intact and if they can be stimulated through hormones. This is where physiological methods come in to clarify if an epithelium has developed proper sealing functions and if vectorial transport works from luminal to basal or *vice versa*.

Special transporting epithelia are not only found in the kidney. Nearly all exocrine gland openings possess a downstream epithelium with the ability to modify fluids in the lumen. The product of exocrine glands is excreted into excretion ducts. In salivary glands, the secretion tubes have the ability to modify osmolarity and ion composition of the secretion through specific membrane molecules in the epithelium. Another example for membrane molecules is found in enterocytes at the intestinal surface. The transporter molecules in epithelial cells bear certain similarities – so we find again aquaporins and epithelial sodium channels (ENaC), which are involved in the solidification of the intestinal content. When such epithelial structures are generated, their typical physiological transport function must, of course, be verified.

[Search criteria: membrane proteins channels transporters]

8.7.5
Receptors and Signals

Many cellular functions are triggered by hormones and a cell needs receptors for these processes. These can be cell surface receptors, in the case of peptide hormones, or intracellular receptor proteins for steroid hormones, whereas other cell functions may be regulated via extracellular electrolytes, such as calcium. In this case, ion channels and ion pumps have an important role to play in the signaling process. Cell surface receptors mostly bind to hydrophilic ligands, which, in turn, may be functionally linked to ion channels or the regulatory proteins of adenylate cyclase. When they are activated, signal molecules may change the functioning of a cell within minutes or even seconds. By contrast, intracellular receptors mostly bind to hydrophobic ligands such as cortisone, cortisol or aldosterone, which activate the transcription process after several hours and induce permanent protein expression for several days. In tissue constructs, however, the production of receptors is often reduced and their connection with cellular reaction cascades is often faulty.

When tissue is generated, one should keep in mind that hormones cannot be added to growth medium because of their poor solubility. They may also bind non-specifically to the culture container surfaces and be absorbed by scaffold material, which would make them unavailable for stimulation. Anyhow, it is a good idea to measure the bioavailability of a specific hormone in order to have an idea of the actual quantity of the molecule dissolved in the medium. Under culture conditions, hormones must often be added in hyperphysiological concentrations in order to ensure their bioavailability.

A peptide hormone such as vasopressin stimulates adenylate cyclase in the renal collecting duct epithelium, ensuring that after the release of the hormone, enough water is absorbed into the body and not too much is excreted as urine. Stimulated by the hormone, the cytoplasm of collecting duct epithelial cells produces cyclic adenosine monophosphate (cAMP), which acts as mediator in the signaling chain. In an adult kidney, adding vasopressin leads to a 30-fold stimulation of cAMP production, whereas in primary cultures of kidney cells, only two- or three-fold stimulation could be achieved. These experiments showed that the vasopressin receptor has been expressed in cultured epithelium. Furthermore, non-specific stimulation of adenylate cyclase by pertussis toxins could be observed. What has not been found yet is an intact signal transduction system including the regulatory subunits of adenylate cyclase. As these examples demonstrate, it has not been possible to increase the level of water transportation, as adenylate cyclase and its regulatory subunits could not be sufficiently stimulated in cultured epithelium.

The successful detection of receptors or regulatory/transduction molecules through immunohistochemical methods or Western blotting can give no more than an indication of the regulating potential of a cell and the expression of receptors alone does not even give away the differentiation status of a cell, as long as no intact signaling cascades have been detected. Receptors in cultured tissue must be able to be stimulated and trigger the natural signaling cascades, otherwise dedifferentiation is a problem.

Another example of the interaction between ligand and signaling effect is the transmitter-controlled ion channel function at synaptic connections. An action potential running along an axon reaches the presynaptic membrane that is separated by a gap from the postsynaptic membrane. The action potential releases a transmitter on the presynaptic membrane. The transmitters bind to receptors on the postsynaptic membrane, which opens up ion channels. The thus released action potential ensures that the stimulus is transmitted. Neurons whose main function is to pass on neural impulses must be in permanent contact with receptors in other tissues in order to fulfill their function. This requires the coupling of receptor and subsequent reaction cascade – a property that must be thoroughly tested in generated tissue. This can only be done through function studies, such as electrophysiological measurements. Experimental neuronal implants into the spinal cords of rats have shown that axon binding does not happen automatically and faultlessly. Axons and dendrites quite often do not grow sufficiently, which leads to a faulty or even non-existent coupling of functions.

[Search criteria: membrane receptors cell signaling]

8.7.6
Cell Surface

The glycocalyx is a layer of oligosaccharides that are bound to extracellular domains of membrane proteins and membrane lipids. Such sugar residues are found on a surprisingly large number of membrane proteins. For example, a water channel protein such as aquaporin 2 is strongly glycosylated, as can be shown through gel electrophoresis and Western blotting. Again, there is a wide variety in the oligosaccharide patterns found in cells and tissues. This can be harnessed for analytical purposes. Lectins bind to terminal sugar residues of membrane structures in animal cells and can be recognized under a light microscope, if coupled with suitable fluorochromes. Thus, lectins are an alternative to antibodies as handy markers in the phenotyping process of cultured cells.

[Search criteria: glycocalyx saccharides lectins]

8.7.7
Constitutive and Facultative Properties

Immunohistochemical analysis and microscopic control are indispensable where exact phenotyping of generated tissue is called for. Proteins should not only be located within the cell, but they must also be examined if they have retained their natural expression pattern. To that purpose, two-dimensional gel electrophoreses of the tissue structures are prepared. The protein spots on the gel plate are then transferred onto nitrocellulose and analyzed using the corresponding antibodies. This method can clarify beyond doubt if a protein spot reacts with the right antibody and if the reacting spot is in its right position on the acrylamide substrate, as far as its charge and its molecular mass are concerned.

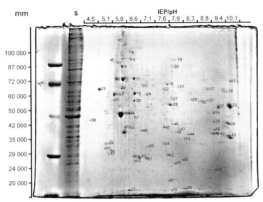

Fig. 8.15: Two-dimensional electrophoresis of a kidney tissue construct for the identification of tissue-specific proteins. A mixture of proteins is separated – first in a pH gradient (horizontally) and then according to molecular mass (vertically). Numerous protein spots, marked with Coomassie blue, can be recognized. These spot patterns are tissue-specific, and individual proteins can be identified by their isoelectric point and their molecular mass.

Before electrophoresis can begin, the tissue constructs must be homogenized and the proteins dissolved. As many of the proteins found in tissue are difficult to dissolve, the sample is prepared in a buffer – usually containing urea and detergents, such as CHAPS, Nonidet or Triton. Then the concentration of the dissolved proteins is determined. About 50 µg of protein are processed for analysis.

In the first dimension, a pH gradient is built up and ampholytes are used to separate the proteins at their isoelectric point. For the second dimension, the focused gels are equilibrated in SDS buffer and separated in according to their molecular mass. The proteins do not appear as bands now, but as roundish spots (Fig. 8.15) that can be detected with markers – named here in the order of increasing sensitivity – such as Coomassie blue, silver staining or fluorescent dyes.

The position of a protein spot in the two-dimensional gel image is determined by its isoelectric point and its molecular mass. These two criteria allow a protein to be clearly identified – including its molecular structure (Fig. 8.15). However, the most sensitive and – above all, most specific – method of identifying a protein spot is Western blotting obtained after running a two-dimensional gel. The use of an antibody will establish beyond doubt whether a spot is specifically recognized or does not react. The sensitivity of this method is so high that about five molecules of a protein in a cell can be identified.

When tissue is being generated, it is usually unclear if the growing conditions can still be called physiological or a stress situation has been created. Here, too, two-dimensional gel electrophoresis and subsequent Western blotting is recommended. In this case, the antibodies used have the ability to recognize constitutive as well as facultatively expressed proteins (Fig. 8.16). Cells will always express their constitutive proteins, whereas facultative proteins are only expressed under specific conditions, such as hormone treatment or stress. In the example given, cyclooxygenases 1 and 2 in renal collecting duct epithelium are shown after exposure to a higher level of NaCl. Alternatively, antibodies to heat shock proteins (HSPs) can be used, as these proteins are only expressed in particularly stressed tissue.

Two-dimensional electrophoresis opens up additional possibilities. Comparative analyses can show that the spot pattern of cultures is not identical with native tissue

Fig. 8.16: Two-dimensional electrophoresis with subsequent Western blotting in a renal epithelial construct. Enzymes are either produced on a permanent base (constitutive enzymes) such as cyclooxygenase 1 (A) or only in certain stress situations, as in the case of cyclooxygenase 2 (B), a facultative enzyme expressed under excessive exposure to NaCl.

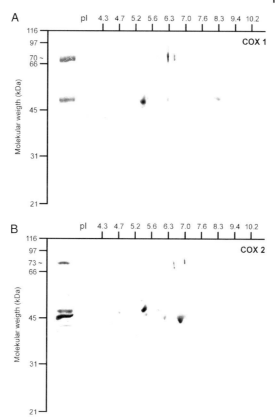

and that spots are found in atypical positions. Such a result provides an opportunity to detect changes in tissue quality and beginning cell dedifferentiation. The suspicious spot is punched out of the gel. The protein then undergoes sequence analysis (MALDI-TOF), which will give part of the amino acid sequence. Comparison with a sequence database will show up overlaps with known proteins and give indications about its whereabouts and its possible function.

[Search criteria: atypical protein expression cell culture]

8.7.8
Detection of Tissue Functions

The analysis of tissue differentiation must not only take morphological, immunological and biochemical aspects into account, but also functional or physiological parameters. In the water or electrolyte-resorbing tubular structures of the kidney, type 2 aquaporins are not only found in the luminal plasma membrane, but also in vesicles within the apical cytoplasm. Immunohistochemical analysis does not only pick up a fluorescence signal limited to the luminal plasma membrane, but also a diffuse reaction in the apical cytoplasm. We know that water resorption is regulated by the hormone vasopressin. If vasopressin is added to the culture medium on the basal side of functionally intact cultured epithelium, the natural cascade of reactions should be triggered. Vasopressin first binds to the V2 receptor. The aquaporins located in the vesicle fuse with the luminal plasma membrane. This channel translocation could be detected immunohistochemically through the disappearance of the diffuse signal in the apical cytoplasm, while the signal in the luminal plasma membrane has become very distinct, due to the integration of water channels. In this tissue-specific example, it is possible to analyze receptor expression, the development of signal transduction and the embedding of channel structures into vesicles with membrane structures – all of which indicate that the tissue produced is functionally intact.

It must be said that in spite of all our culture experiments, we have not been successful yet in producing a perfect water-transporting epithelium, and all the tight junctions and the presence of the necessary channel structures – including their receptors – we are still far off the mark. For reasons still unknown, the reaction cascade between the vasopressin receptor and the adenylate cyclase involved was not fully developed.

By contrast, Na^+ transport, which can be stimulated by aldosterone and inhibited by amiloride, developed perfectly in generated kidney collecting duct epithelium. In a model developed by another research group, it could be observed that in cultured tissue, the channel structures typically found in the apex only appeared in an atypical position – the basolateral plasma membrane. This means a partial reversal of cellular polarization has taken place and, despite numerous efforts, no satisfying solution has been found to solve the problem.

Isolated cultured neurons are frequently used as models. For the purposes of tissue engineering, however, where the focus is on the functioning of generated tissue, the transmission of neural impulses from one cell another must be continuous. This is indispensable for the repair of defective integrative processes within the neural system. In Parkinson's disease, for example, dopamine synthesis has been disrupted in the midbrain. In healthy humans, the transmitter dopamine couples the mesoencephalic neural tissue to the basal ganglia, thus coordinating the motor system. When treating the diseased brain area with cell or tissue implants, at least two conditions must be met. First, the construct should have only limited mitotic capacity, and the cells should not migrate. Second, the transplanted cells should be dopaminergic, i.e. capable of synthesizing dopamine. It cannot be assumed that after implantation of a tissue construct or of isolated cells into the region, dopamine synthesis will automatically resume. This is why experimental evidence must show beyond doubt that the neurons do in fact integrate into their future tissue environment, that they are able to produce the required transmitter over a long period of time and process it at the synapses.

The produced tissue must be medically functional, be useful to the patient and not cause any harm, and it is unimportant if the tissue used for the implant is embryonic, semi-mature or differentiated tissue. The same applies to an extracorporeal module in support of a liver or kidney function that is brought into contact with a patient's blood filtrate in order to metabolize or dispose of certain substances. Modules with gener-

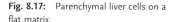

Fig. 8.17: Parenchymal liver cells on a flat matrix.

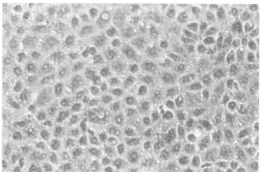

ated epithelia should be able to take over such functions. What use is, for example, a cultured epithelium produced for a kidney module, if an aquaporin 2 channel has been established in the luminal plasma membrane, but the water that has been resorbed cannot find an exit at the basal membrane because aquaporins 3 and 4 have not been expressed, due to cell dedifferentiation? In addition, the generated tissue must show a high degree of resistance to the rheological, hydrostatic and pulsatile stress caused by flowing serum or medium. In particular, it should be resistant to the concentrations of urea in the serum. To our knowledge, systematic studies of these aspects have been few and far between.

Similar problems had to be overcome when setting up extracorporeal liver modules (Fig. 8.17). Before these are tried out on a patient, they must be in a stable stand-by condition for weeks and months. When needed, the cultured cells must be able to deliver instantaneous and sufficient detoxification. Over a longer period of time, it must be possible to regulate the synthesis of blood serum proteins, e.g. to compensate a lack of blood clotting factors. All these problems must be experimentally investigated in depth before we can even think of general applications of liver modules.

When it comes to the application of tissue engineering in the treatment of patients, we are no longer looking at small-scale analysis in the laboratory, but at a larger production scale. Experience has shown that problems that seemed to have been solved on a small scale may crop up again in the scaling-up process. It cannot be taken for granted that cells intended for larger module projects can simply be grown in cultures and will maintain their degree of differentiation and functionality – so crucial in organ modules. A large number of factors are involved – some of which can be experimentally controlled, while others are still completely unknown.

[Search criteria: differentiation dedifferentiation functional expression]

8.8
Quality Assurance

Many teams and companies have sprung up that produce tissue constructs for clinical treatment on a commercial basis. For example, a lot of work is going into orthopedic implants – growing chondrocytes on a wide variety of scaffolds in order to produce resilient cartilage tissue. Such tissue could be implanted into damaged joints. So far, it remains unclear which strategy or method is best to obtain the desired result. The same applies to teams that are working on the production of typical bone, tendon and loose connective tissue. Many teams are using heart muscle cells, glandular tissue, endocrine tissue or neuronal cells in order to produce functional tissue. Again, no comparative analytical data are available that look at the quality of the constructs generated. It is often said that the best scaffold will yield the best construct – a statement that still awaits confirmation by verifiable experimental evidence.

When looking at current publications of competing teams that produce identical or similar tissue constructs, the difficulty lies in deciding which construct exhibits the highest degree of tissue-typical differentiation – so many different methods and analytical procedures are being used. In addition, the markers used to identify differentiation characteristics vary widely, making it next to impossible to verify and compare the quality claims for the various tissue constructs in the absence of universally valid criteria. A comparison is not made any easier by the fact that many of the antibodies that have been used for the determination of tissue differentiation are often not available to other teams or do not yield the result described, as we often found.

For its biomedical application, the decisive factor is not only the proliferation of cells and tissue generation under optimum culture conditions, but, above all, the quality of the final construct. This includes that the procedures involved are reproducible in the laboratory, as postulated in GMP guidelines. These guidelines define all lab procedures and their documentation. Strangely enough, the final product, i.e. the cells and tissue generated, are not mentioned. Such an essential aspect of objective quality assessment must no longer be ignored. After all, the all-decisive question – how suitable a construct is for medical treatment – depends on the degree of cellular differentiation and the absence of atypical characteristics.

The aspect of optimal tissue development becomes even more important where stem cells are involved, which seem to be favored for future tissue production. Independent of the source of the stem cells – which may be embryos, umbilical cord blood or adult tissue – it is a case of proliferating embryonic cells of the same type and transforming them into various tissue cells with the help of suitable morphogens and growth factors. These cells, too, must be grown on a scaffold, as described above, in order to mature into functional tissue. It is crucial in this context to be able to analyze first if the cells have developed into uniform or diverse types of tissue cells.

Current data show that precursors of fat, cartilage and bone tissue can all develop from the same stem cell line (Tab. 8.2). It would therefore be appropriate to analyze what makes up the differences between all these precursor cells deriving from the same cell type. In a next step, it must be clarified if all stem cells developed into adi-

Tab. 8.2: Example of a rather superficial identification of lipoblasts, chondroblasts and osteoblasts that may develop from a stem cell.

Tissue	Cellular production	Histology/immunohistology
Fat	Lipid droplets	Oil red staining
Cartilage	Sulfated proteoglycans; collagen type II synthesis	Alcian blue staining (pH 1); collagen type II antibody
Bone	Alkaline phosphatase (AP); calcification	Histochemical: AP; Von Kossa staining

poblasts, chondroblasts and osteoblasts or if a certain percentage of cells evaded this developmental step, retaining their stem cell properties. After implantation, this population would behave like stem cells and not like mature tissue cells. Such non-developed or possibly partially developed tissue cells could pose a risk. After implantation of the construct, these cells might migrate into host tissue, forming ectopic tissue structures or even tumors.

It is also possible, however, that only part of the cells will develop into adipoblasts, while others are developing the properties of chondroblasts and osteoblasts. These cells must be identified and eliminated if functional tissue is to develop. If heterogeneous tissue – containing various cell types that have not been recognized – were implanted, the risk involved would be incalculable.

[Search criteria: tissue markers differentiation histochemistry]

8.8.1
Appearance of the Construct

The stem cells in which differentiation has been stimulated must be transformed from a monolayer growing at the bottom of a culture dish into three-dimensional functional tissue. This involves establishing them on a scaffold. If cartilage tissue is to be generated, just analyzing if the cells growing on the scaffold produce collagen type II is not enough (Tab. 8.2). It is equally important to know if the type II collagen is polymerized in the ECM and if it develops into a physically resilient intracellular substance. In fat tissue, an additional question to be answered is whether reticular fibers are developing that form a three-dimensional network to stabilize the cell.

Staining with oil red will only detect those lipid-containing cells that grow as a monolayer at the bottom of a culture dish. As these cells still lack many characteristics, we can be certain that they are not mature adipocytes. If only identified by staining, they could – in theory – even be steroid hormone-producing cells. Similarly, in muscle cells, from the expression of myosin alone, no conclusions can be drawn on the long-term contractibility of the cells. Most significantly, it is not even clear if skeletal muscle fibers are produced instead of heart muscle cells.

A wide range of antibodies is now commercially available that can identify the basic properties of any tissue beyond reasonable doubt (Tab. 8.3). If such antibodies could also be used to clearly identify the properties of generated tissue, which would then

Tab. 8.3: Examples of tissue-specific proteins in an adult organism, which can be clearly identified immunohistochemically or by Western blotting.

Tissue	Marker
Connective tissue	Collagens, Vimentin
Epithelial tissue	Cytokeratins, Occludins
Muscle tissue	Desmin, Myosin
Neural tissue	Neurofilaments, Myelin

have to be compared to reference tissue, it would be a step in the right direction. In the field of pathology, it is already common practice to diagnose diseases and tumors with the help of a wide range of markers.

[Search criteria: markers differentiation dedifferentiation control]

8.8.2
Analytical Microscopy

The distribution pattern of cells on or in a scaffold can be shown very easily, using a fluorescent dye such as DAPI, which reacts to constituents of the nucleus. By looking at the distribution of nuclei under a fluorescent microscope, it should be possible to tell at the first glance whether the biomaterial used is evenly or only partially populated with cells. The additional use of gold or fluorescent marked antibodies should give additional information about the presence of tissue-characteristic ECM proteins and to what degree the cells have been able to develop certain differentiation markers (Fig. 8.18).

Despite all efforts to optimize culturing methods, it cannot be excluded that the tissue cells that grow on the scaffold will only develop a limited range of their natural differentiation profile. While it is fairly easy to assess the scope of such a development under a light microscope, it is far more time consuming, if not difficult, to find its

Fig. 8.18: Clear immunohistochemical identification of renin-producing cells. Three immune-positive cells growing on a layer of unmarked cells are shown.

causes. It is best to examine the tissue construct concerned carefully under an electron microscope in order to assess possible consequences.

The preparation of a tissue construct on an artificial scaffold for electron microscopy (especially TEM) is usually very labor intensive, as it requires fixing the constructs with greatest possible care to avoid shrinking or tearing. Then, the samples need to be dehydrated for further processing, which is usually done with solvents such as alcohol or acetone. While the dehydration of the tissue itself is straightforward, major problems may arise from the scaffold material contained in the construct. If it is not solvent resistant, it may easily be damaged or even dissolve in a rising solvent gradient. A way has to be found that will preserve the tissue in the electron microscopy sample, but avoids damaging the scaffold used.

When fixing tissue constructs, great care must be taken to protect them from osmotic shock, which could lead to changes in the ultrastructure. The fixation is carried out under iso-osmotic conditions, using a defined concentration of glutaraldehyde, especially purified for use in electron microscopy. This is a very straightforward procedure that ensures maximum tissue protection. A small piece of tissue is put in a culture dish containing, for example, precisely 1 ml of medium without serum or protein additives. In a separate tube, a solution is prepared containing the same culture medium plus 3% glutaraldehyde. Precisely 1 ml solution is taken and pipetted into the 1 ml medium with the tissue sample. This results in 2 ml of solution containing 1.5% glutaraldehyde. This seems to be the most tissue-protective method of fixation currently known.

It is also important to make sure that the fixation steps are carried out on a cooling surface or in an ice bath at around 2 °C. The fixed samples are then transferred into PBS which contains calcium and magnesium, 0.1 M sodium cacodylate as well as 0.1 M sucrose and has a pH value of 7.4. The samples can only be kept in this solution for a few days in the refrigerator. Postfixing takes 60 min and is done with a solution of 1% osmium, 0.1 M sodium cacodylate and 0.1 M sucrose at a pH of 7.4. The samples are then stored in the refrigerator and washed several times in PBS until the supernatant becomes clear. The postfixed tissue can be stored in the refrigerator indefinitely.

The samples that have been postfixed with osmium must be dehydrated before they can be embedded in epoxy resin. Depending on tissue and scaffold material, a balance must be struck between optimum dehydration of the tissue and minimal damage to the scaffold by the solvent. It is also necessary that the dehydrated samples are well penetrated by the epoxy resin in order to achieve homogenous polymerization. It depends on the solvent resistance of the scaffold which of the following dehydrants should be used in a rising gradient – ethanol or butanol, and if acetone, propylene oxide or some other intermedium should be used before the sample can be embedded.

Another question that may cause headaches, but is often underestimated, is the right choice of epoxy resin for the embedding of the tissue block. Whether Spurr, Araldite or Epon are used as embedding resin must be decided on the basis of the tissue material in hand. During the fixation, dehydration and embedding processes, the tissues develop varying degrees of brittleness, which affect their behavior when the block is cut using a glass or diamond knife in the ultramicrotome for ultrathin sections. The problem is compounded by the fact that in cultured constructs, not only the tissue must be cut, but also the scaffold. It is often found that while the tissue itself can be cut quite easily, the scaffold tends to crack and splinter, making it thus impossible to obtain usable sections. In such a case, one would have to start all over again – experimenting with different fixing solutions, new dehydration gradients and finding a more suitable epoxy resin embedding material. While the optimizing procedure may prove unexpectedly labor-intensive, the essential findings they yield could not be obtained any other way.

If a particular antigen has to be located in the ultrathin section, using its antibody, electron microscopic work becomes even more complicated. In that case, the fixation process must be extremely gentle and the tissue construct must be embedded in a special water-containing resin in a refrigerated atmosphere. Only after incubation of the ultrathin sections will it become evident if the embedding procedure was gentle enough not to damage the structure of the antigen and only then can it be detected (Fig. 8.19). If these especially prepared sections do not yield an antibody reaction, a more tedious procedure lies ahead – pre-embedding incubation of antibody and the preparation of ultrathin sections of frozen material. Here again, only the experiment will show if the scaffold used can be cut in a frozen stage.

The processing of tissue constructs for analysis under a scanning electron microscope is relatively easy in comparison. The samples that have been fixed in the way described above must first be dehydrated. PBS is replaced by distilled water, and washed for 10 min each time in 35, 70, 85, 95 and absolute alcohol. The samples can be dried using a critical point drying (CPD) apparatus or, alternatively, samples can be dipped into a small amount of hexamethylsilane. The hexamethylsilane is left to evaporate under a hood; no further apparatus is needed. The sample is stuck to a holder with double-sided tape. In order to create electrical conductivity in the tissue, colloidal silver must be spread between the edges of the tissue and the holder. A thin layer of gold or carbon is then vaporized onto the tissue. These steps take several hours until, eventually, the samples can be analyzed in the scanning electron microscope.

Fig. 8.19: Immunohistochemical analysis of an ultrathin electron microscopic section. The black gold granule markings in the apical plasma membrane and in the cytoplasm of generated collecting duct epithelium are clearly visible.

[Search criteria: electron microscope analysis cell culture]

8.8.3
Detection of Tissue Structures

When two different types of cells are to be compared, antibodies could be used to show if the cytoskeletal structures that have developed are the same or not. If, however, tissues are to be compared, it will not be sufficient to compare cellular elements only, as extracellular components come into the equation (Fig. 8.20). Therefore, it makes little sense to use a marker for cellular properties in one tissue and then compare it to a marker for ECM in the other. In profiling, a clear distinction must be made between (a) the properties of cells and (b) the properties of tissue.

In an organism, embryonic stem cells, mesenchymal progenitor cells and immature tissue cells go through a long chain of intermediate stages before they mature into functional tissue (Fig. 8.2). The more complex the developmental chain, the larger the scope for faulty development in cultured tissue. Chances are that the desired type of cells may develop, but then they may not and the cells developing in the construct could be very different indeed. This makes state-of-the-art immunohistochemical typing indispensable, if the quality of the tissue construct generated is to meet objective criteria (Tab. 8.4).

Markers such as CD14 (monocytes, macrophages), CD45 (leukocyte antigen) and CD34 (stem cells, progenitor cells) can show if, for example, blood cells have developed in stem cell cultures. The presence of endothelial cells can be demonstrated using MUC 18 and vascular cell adhesion molecule 1 (VCAM-1). Antibodies to neurofilaments detect properties in neural cells. The presence of smooth muscle cells can be established using MyoD from myocytes and α-smooth muscle actin. The properties of cartilage can be identified through the presence of type II collagen and proteoglycans such as aggrecan and chondronectin, while the characteristics of bone can be established by analyzing alkaline phosphatases, type I collagen, osteonectin, osteopontin, osteocalcin and bone sialoprotein (BSP), thus distinguishing it from cartilage tissue. Fat tissue can be identified by the detection of peroxisomal proliferation activated receptor γ2 (PPARγ) and reticulin. Fibroblasts can be recognized by detecting type III collagen and fibroblast growth factor 2.

This considerable list is by no means complete, but it gives an insight into the wide range of possibilities of cell biological tissue typing at a transcriptional as well as at a translational level. What we want to show is a method that can recognize uniform as well as diverting cell and tissue development, thus making it possible to distinguish

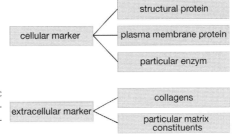

Fig. 8.20: Identification of tissue-characteristic properties using cell biological methods. The distinction between cellular and extracellular markers is consistent.

Tab. 8.4: Immunohistochemical typing schema for the classification of cells that are developing form embryonic or fetal cells into functional tissue. Profiling identifies identical, similar and – particularly – diverse structures.

Cell type	Marker
Hematopoetic cells	CD14 (monocytes, macrophages)
	CD45 (leukocyte antigen)
	CD34 (stem cells, progenitor cells)
Endothelium	MUC 18
	VCAM-1
Neuronal cells	Neurofilaments
Smooth muscle cells	MyoD
	α-smooth muscle actin (α-SM actin)
Cartilage	Collagen type II
	Proteoglycan
	Chondronectin
Bone	Alkaline phosphatase
	Collagen type I
	Osteonectin
	Osteopontin
	Osteocalcin
	BSP
Fat	PPARγ
	Reticulin
Fibroblasts	Collagen type III
	FGF2

between desired and undesired tissue developments at an early stage. Missing characteristics and possibly atypical structures can be reliably analyzed. Whenever an experiment is set up, a decision must be taken on the merits of the case if it is sufficient to detect structural molecules only or showing the intracellular functional cascades though Western blotting with appropriate antibodies would be preferable. Sets of suitable antibodies, providing the all-important objective measuring criteria of differentiation, are now readily available from suppliers.

[Search criteria: immunohistochemical markers tissue differentiation]

8.8.4
Definitive Recognition of Maturation

In order to proceed with quality assurance, it is important to determine the maturation status of the cells in the tissue construct beyond doubt. It should be established if the cells have reached an adult stage or if they still retain some embryonic or semi-mature properties. Morphological methods alone will be not sufficient to provide an answer to these pressing questions – a whole range of cell biological techniques is called for. At a transcriptional level, current gene activity can be measured, whereas at a translational level, protein expression can be observed. It is vital to find out if the developing tissue cells have mixed characteristics or form atypical proteins.

Not only the upregulation of properties could be tested in artificial tissue, but also structures that are found in embryonic and maturing cells, but are lost in adult cells (Tab. 8.5). In their development from the embryonic stage to the terminal, differentiated stage, tissue cells go through many intermediate stages during which new properties are acquired, while others are lost. Fetal liver cells, for example, produce fetoprotein, but stop producing it with increasing differentiation and functionality. Similar observations could be made regarding carcinoembryonic antigen (CEA). $P_{CD}Amp\ 1$ is another antigen found only in embryonic or maturing renal collecting duct epithelial cells, but not in mature cells. Unfortunately, only very few examples of downregulation of properties are known so far and, accordingly, very few markers.

Tab. 8.5 Examples of the identification of proteins that help recognize a clear loss of embryonic properties in tissue.

Embryonic	Adult
α-Fetoprotein	–
CEA	–
$P_{CD}Amp1$	–

[Search criteria: embryonic tissue development transient protein expression]

8.8.5
Transitory Expression

In an ideal world, it would be possible to stimulate isolated tissue cells or stem cells with more or less embryonic properties that they develop into functional tissue cells complete with ECM, just as they would differentiate *in vivo*. The generation of such constructs under *in vitro* conditions is the result of many very complex developing mechanisms that take a surprisingly long time – several weeks. During this time, immature cells develop into tissue with more or less specific properties, going through several intermediate stages on the way.

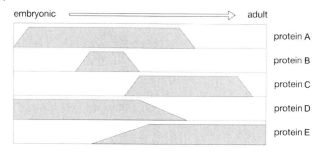

embryonic ⟹ adult

protein A

protein B

protein C

protein D

protein E

Fig. 8.21: Schematic representation of transitory expression phases during tissue development. Protein A is produced over a long period during the embryonic stage, while protein B is only briefly expressed. Protein C only emerges from the late embryonic stage through to adulthood. Protein D is downregulated during the upregulation of protein C. Protein E only appears when adult structures are formed.

Proteins that have been observed during the various stages of development along the embryonic and fetal time axis can be expressed in various ways (Fig. 8.21). In analogy to the development of organisms, generated tissue goes through a phase from a totally immature (embryonic) to a functional adult stage. In this process, proteins are not simply up- or downregulated, but there are also varying transitory expression phases. This means that the upregulation of proteins can happen simultaneously or lag behind the transitory presence of proteins.

There are only very few commercially available markers that detect these transitorily expressed proteins in specific tissues. Thus, there is nothing for it but to home-produce tissue-specific markers that show the developmental status of maturing cells. These antibodies can be used to find out if specific properties are developing, at what stage of development upregulation sets in and if atypical proteins are present in the tissue. In the future, standard antibodies that could be available to all research teams interested could back up such antibody reactions.

[Search criteria: embryonic development temporal transient expression]

8.8.6
Making New Markers Available

While commercially available antibodies are good detectors of basic properties in tissue, there is a shortage of markers that help distinguish embryonic structures from semi-mature and functional stages. In order to answer these specific questions, it is well worth producing monoclonal antibodies that can be isolated after immunization with embryonic, semi-matured and differentiated tissue.

Antibodies are globular proteins (immunoglobulins) that are produced and excreted by B-lymphocytes in response to the presence of a foreign substance, an antigen. A B-lymphocyte can only recognize a specific antigen and produce only one kind of antibody to this particular antigen. This specific response is utilized in research to detect specific molecules and make them visible. An antibody has a specific affinity to a specific place on the antigen, which is called an epitope. An antigen may possess several different epitopes to which several different antibodies can bind.

In an animal organism, immunization through an antigen is always followed by the activation of a variety of immune cells for the production of antibodies. This heterogeneous immune response gives rise to various clones that produce various antibodies. These polyclonal antibodies are all directed towards the same antigen, but since they do not derive from the same mother cell, their structures may vary and they may bind to different epitopes on the antigen.

In practical applications, monoclonal antibodies are given preference over polyclonal antibodies – mainly because their structure and function can be precisely defined and standardized, and they can be produced in almost indefinite quantities. In 1975, Cesar Milstain and George Koehler developed a method of producing vast amounts of monoclonal antibodies, based on the principle of artificial fusion of tumor cells (myeloma cells) and antibody-producing B-lymphocytes (mouse, rat, rabbit, guinea pig, human).

The fusion products from myeloma cells and antibody-producing B-lymphocytes are called hybridoma cells and combine the useful properties of both parent cells, including continuous growth, production of specific antibodies and specific enzymes.

To obtain monoclonal antibodies, the cells must be cloned. This can be done through a variety of procedures, such as the limiting dilution method. Dilutions are prepared of the cell suspension to be cloned – containing about 5, 1 and 0.2 cells/ml. These diluted cell suspensions are plated into special dishes and cultured. In order to make sure pure clones have been isolated, the cloning process must be carried out a second time. The growth of the young hybrids is stimulated by adding to the culture medium freshly isolated splenic or peritoneal cells as feeder cells. By their presence and their secretion of natural growth factors, these cells create a stimulating environment for the hybridoma cells. Not every fusion of a myeloma cell and a B-lymphocyte results in a useful antibody-producing hybrid. This is why the culture supernatant in the dishes containing hybrid cell clones must be tested to see if there is a reaction to the antigen that had been used for immunization in the first place.

To obtain an antibody to a specific protein, it is not always necessary nowadays to immunize an animal. The same effect can be achieved by *in vitro* immunization, which involves isolating and culturing splenic cells. The protein to which an antibody is to be produced is added to the culture medium. After 3 days of culturing, the splenic cells are fused with myeloma cells to produce hybridoma cells. Already after 10–14 days, a test will show if the hybridomas are secreting specific antibodies into the culture medium. This method is unequalled in its rapidity and reliability.

It is best to subject the many antibodies created to immunohistochemical tests, using frozen sections of embryonic, semi-matured and adult tissues (Fig. 8.22). The fluorescing binding signal of a created antibody may indicate, for example, that only embryonic, but not adult, structures have been recognized. Inversely, antibodies can be obtained that mark adult, but not embryonic, cells. Possibly, other antibodies may be found that recognize various stages of intermediate cell development.

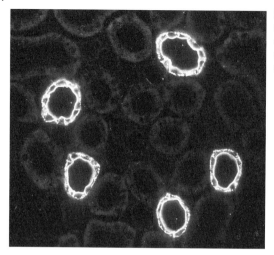

Fig. 8.22: Binding of fluorochrome-marked antibodies to specific tubular cells of the kidney. Such markers can surely identify individual cells in tissue constructs.

The technique described can be applied to all tissue. In addition to the immuno-histochemical test, the newly identified protein could be isolated through Western blotting of a two-dimensional electrophoresis and made visible using the newly generated antibody. The protein is exclusively cut out of the gel and undergoes microsequencing (MALDI-TOF). The amino acid sequence found permits an identification of the protein. Often, this can give clues whether the generated antibody is recognizing a functional or a structural protein and in which cellular structure it can be found. It is very likely that this method will help find many proteins that have been given very little or no attention, but could become important differentiation markers in the future.

An excellent overview of suppliers of antibodies in tissue differentiation can be obtained from *Linscott's Directory of Immunological and Biological Reagents* (http://www.linscottsdirectory.com).

[Search criteria: specific production monoclonal antibodies hybridoma]

8.9
Implant – Host Interaction

Tissue constructs are created for medical use. Let us look at a typical example – the implantation of artificial cartilage tissue into a damaged joint surface. First of all, the knee joint must be opened to insert the generated construct. This can be done either by open surgery or minimally invasive techniques.

In classical open surgery, any shape of tissue construct on a rigid matrix could be inserted into the damaged surface (Fig. 8.23). This surgical technique has the advantage of allowing pre-matured and physically resilient constructs to be implanted. Its drawback is its longer healing time and prolonged stay in hospital.

If minimally invasive techniques are used, a construct can only be implanted if it is small and, above all, flexible enough (Fig. 8.23). It is inserted into the knee through a

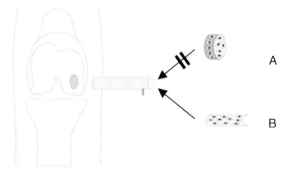

Fig. 8.23: Minimally invasive surgery on damaged cartilage in a knee injury. A rigid construct (A) cannot be inserted into the knee through the minimal diameter of the surgical instruments, while a flexible populated matrix (B) can be rolled up and pushed through.

channel with an inner diameter of a few millimeters in order to be attached to the damaged cartilage surface. This is technically feasible, the stay in hospital is short and the wounds usually heal quickly. From a cell biological view, only flexible matrices can be used; in other words, during the *in vitro* developing process, no rigid cartilage matrix must form. Thus, only tissue precursors can be implanted which are more or less immature and have not yet developed functional properties.

After implantation, the tissue construct must grow into an integral part of its new environment. Ideally, the implant would now develop a resilient surface. In order to grow in, it must keep very close contact to the surrounding tissue on the basal and

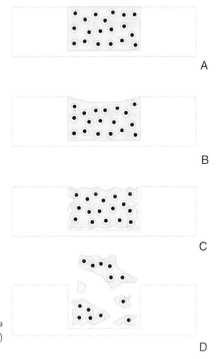

Fig. 8.24: Accuracy of fit after implantation, e.g. to a joint surface: (A) perfect fit, (B) surface alteration, (C) shrinking and (D) loss.

lateral side. Various experiments, however, have shown that this ideal scenario does not often materialize. Changes at the surface and shrinking processes have been observed or the implant is not integrated into the surrounding tissue (Fig. 8.24).

We must keep in mind that implantation of artificial tissue does not always lead to healing, but can also elicit inflammatory and immunological responses. These may be provoked either by the implanted tissue cells or by the scaffold material. There is always a primary reaction at the interface between the implant and the surrounding tissue. If the implant has toxic properties, it will induce necrosis in the immediate neighborhood. If an implant is inert, it will not develop any connection to the surrounding area, forming an atypical connective tissue capsule. If, on the other hand, the implant is bioactive, the surrounding tissue will soon integrate it functionally. If the implant even has a biodegradable scaffold, this will be replaced by interactive growing tissue – as long as the scaffold material is replaced gradually and the resorption rate is in line with the regrowth of tissue. Thus, the required resilience of the regenerating tissue can be ensured.

Especially in the regeneration of bone, reactions have been often underestimated. With the use of biotolerant scaffolds, distance osteogenesis has often been observed. The implant is overgrown by a connective tissue layer, which prevents direct contact between the implant and the surrounding bone. Bioinert scaffolds lead to contact osteogenesis. In this case, the implant is not surrounded by connective tissue and osteogenesis takes place in the implant as well as in the bone. However, integration at the implant surface is not always very good.

The best conditions are found in bioactive scaffolds where cells from the immediate surroundings grow immediately into the implant to form functional tissue. This process depends on the osteoconductive and osteoinductive properties of the scaffold. For optimal osteoconduction, the surface of the scaffold must contain chemical and physical properties that support three-dimensional proliferation of tissue. For osteoinduction, the scaffold must have properties that induce differentiation in the proliferated osteoblast precursors. Only then can the progenitor cells develop into osteoblasts and osteocytes and build functional osteons.

[Search criteria: graft host interaction tissue engineering]

9
Perspectives

In general, very little is known about the development of functional tissues with their different terminally differentiated cell types. Precursors of tissue cells develop from embryonic cells that form socially interacting networks in a special matrix in the course of development and finally take on the characteristics of the adult tissues. These processes, which naturally take place in the body, are not directed by an individual growth factor, but by a variety of completely different mechanisms (Fig. 9.1). These include adhesion to the ECM, the control of the cell cycle in mitosis and interphase, the effects of exchange between neighboring cells, the effect of hormones, and

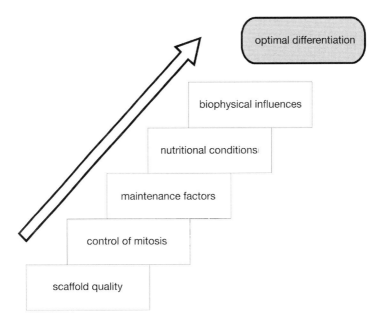

Fig. 9.1: Scheme for the multifactorial development of artificial tissues. Functional tissues can be manufactured under *in vitro* conditions only if optimal scaffolds are used that succeed in controlling the mitosis and interphase periods with optimal hormonal and nutritive influences. In addition, biophysical influences, such as pressure, temperature and rheologic stress, can positively affect differentiation.

Tissue Engineering. Essentials for Daily Laboratory Work W. W. Minuth, R. Strehl, K. Schumacher
Copyright © 2005 WILEY-VCH Verlag GmbH & Co. KGaA, Weinheim
ISBN: 3-527-31186-6

biophysical influences such as pressure, liquid movement, oxygen content and nutrition. How these interactive processes begin and take place in a temporally coordinated manner has hardly been analyzed. Which of these factors is, hierarchically, the most important can only be measured with difficulty. However, it is clear in experimental work with artificial tissues under *in vitro* conditions that none of these influences can be underestimated.

Natural development means reaching an optimal functionality in each of the specific tissues, by which the cells and ECM achieve a typical differentiation. Current experimental data from the production of artificial tissue show that promising beginnings have been made, but the goal of generating optimal functionality in constructs is far from being achieved. These problems can only to be solved by the development of improved culture methods, and appropriate specific scaffolds, microreactors and media, which must be optimally adapted to the particular needs.

It would be ideal if isolated tissues or tissue constructs could be kept for longer periods under culture conditions without the migration, restructuring and dedifferentiation of the cells. Tissue banks could be specifically developed with such constructs. When needed, a tissue implant in the necessary form, with the necessary logistics could be ready in little time.

Using optimal tissue constructs, one could also investigate the emergence of acute and chronic inflammations as well as degenerative illnesses under pure *in vitro* conditions for the first time. One could use these models in order to experimentally setup clearly defined injuries so as to collect information about the regenerational ability of the constructs. In addition, tissues produced in such a way would be ideal models for analyzing the healing of newly developed biomaterials without interference to an organism.

At this point, there is too little knowledge about the interactive processes of the individual tissues. Also, the cultivation of tissues outside of the body has not yet been technically mastered. How does one want to develop optimal constructs when a piece of isolated cartilage or bone in optimal form cannot be kept alive for long under culture conditions? How much better would a module with liver, pancreas or kidney parenchymal cells work if one could only optimize the culture conditions on a small scale before learning their differentiation behavior? With time, one could learn how the differentiation can be optimally directed. It would be fantastic if the targeted growth of new axons could be stimulated in isolated segments of the spinal cord. However, science is not so far advanced that healing paralyzed patients could be envisaged in the immediate future.

We must guard against giving hasty and false promises when it comes to the generation of tissues. In the future, only one thing will help in solving the many cell biological problems with the emergence of tissue constructs – the continuation of intensive research work at the sterile bench. The recognition of the necessity and an internal readiness to investigate the current questions will be required. Solid financial support over many years will be necessary to do this.

[Search criteria: tissue engineering advances review]

10
Ethical Aspects

Inevitable considerations of the ethical aspects arise when artificially manufactured tissue is to be used for implantation. The question at the forefront is of the quality of the construct and the risks it presents. No less important are the considerations about the integration of the constructs. If autologous adult cells are used, few ethical doubts will arise. However, the use of embryonic stem cells or master cell lines is very different. In therapeutic use of embryonic stem cells, it must be clarified whether all implanted cells really develop into the desired tissue. It is possible that a small part of the cells behaves indifferently at first and thus inconspicuously. With time, however, unwanted tissue or even a tumor can develop. This is also a sensitive subject, since the cells of a tissue construct do not only stay within the area of the implantation, but are able to migrate into the whole body. It cannot be stressed enough that we are only at the beginning of fascinating developments in tissue engineering. For this reason, most of the related ethical questions can be only raised at this time, but not answered satisfactorily.

A variety of diagnostic, preparative, analytic and logistic activities are linked to one another in tissue engineering (Fig. 10.1). A very complex process results from this. At each point in this chain, an error can develop that may bring with it serious consequences. For medical and ethical reasons, therefore, errors must be recognized as early and as quickly as possible. For this reason, perfect documentation of all steps within the developing chain of events in a tissue construct becomes of central importance. Only with this critical consciousness can the biomedical risks for the future be

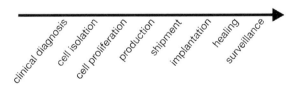

Fig. 10.1: The production of artificial tissue is a multifactorial process in which a variety of different activities come together, e.g. clinical diagnosis, cell isolation, proliferation of the cells, production of constructs, dispatching the constructs, implantation, and the healing and monitoring of the patient over many years.

Tissue Engineering. Essentials for Daily Laboratory Work W. W. Minuth, R. Strehl, K. Schumacher
Copyright © 2005 WILEY-VCH Verlag GmbH & Co. KGaA, Weinheim
ISBN: 3-527-31186-6

minimized. This is, also, certainly the way to approach ethical issues on a purely argumentative basis.

[Search criteria: ethical considerations tissue engineering transplantation]

Glossary

24-well culture plate	Culture plate with 24 single chambers.
2-D electrophoresis	Separation of proteins by isoelectric point and molecular weight.
3T3 cells	Fibroblast cell line.
α-Actinin	Anchoring protein, which connects actin filament to each other.
α-Fetoprotein	Protein found mainly in fetal tissue.
γ-radiation	Electromagnetic radiation emitted during radioactive decay.
Aerobe/anaerobe	Dependent on the presence/absence of oxygen.
Rejection reaction	Reaction of the immune system to foreign material or tissue.
Acetone	Colorless, highly flammable and aromatic-smelling fluid (dimethylketone), used in many paints, lacquers and resins.
Activin	Formerly vegetative factor, growth factor.
Adenylate cyclase	Enzyme which converts ATP to cAMP, mostly receptor associated.
Adhesion molecule	Molecule for connection between adjacent cells (i.e. integrin or cadherin).
Adipocyte	Fat cell.
Affinity	Strength of binding between two molecules at a single binding site.
Agarose	Linear polysaccharide from red algae.
Agglutination	Clotting.
Aggrecan	Proteoglycan which aggregates with hyaluronic acid to form large complexes.
Agrin	Secreted protein which controls formation of synapses of innervating neurons on muscle cells.
Actin	Cytoskeletal protein, basic unit of an actin filament, which together with myosin comprises the contractile unit of muscle cells.
Albumin	Serum protein, with transport, buffering and osmotic functions.
Aldehyde group	[-CH=O].
Aldosterone	Hormone produced in the adrenal glands.
Alecithal	Containing little or no yolk.
Alginate	Polymer isolated from the cell wall of algae.
Aliquot	To divide into portions.
Alkalosis	Increase in blood pH to over 7.44.
Alveolus	Single sac in the lung; also the pocket in which a tooth is embedded.

Tissue Engineering. Essentials for Daily Laboratory Work W. W. Minuth, R. Strehl, K. Schumacher
Copyright © 2005 WILEY-VCH Verlag GmbH & Co. KGaA, Weinheim
ISBN: 3-527-31186-6

Alzheimer disease	Pre-senile dementia as a result of progressing atrophy of the cortex.
American Type Culture Collection (ATCC)	Worldwide, the largest collection of deep-frozen cells.
Aminopterin	Inhibitor in the DNA synthesis pathway.
Amino acid	Organic, nitrogen-containing molecules that serve as the building blocks of proteins.
Amphibian	An animal, which can live both in and out of water, passes most of its adult life on land, and returns to the water to lay its eggs.
Amphoteric	Having properties of both an acid and a base.
Amphotericin	Strong antimycotic (Fungizone).
Amylase	Enzyme which splits starch and glycogen.
Anatomy	Study of the internal structure of the body, and the placement and structure of organs and tissue.
Anchor proteins	Proteins which help attach cells to surfaces.
Angiogenesis	Formation of new blood vessels during embryonic development or wound healing, as well as in the building of new tissue in solid malignant tumors or during the menstrual cycle.
Angiopoetin	Ligand of Tie-2, affects blood vessel development.
Anion exchanger	Ion exchanger that exchanges ions (cations, positively charged) with counterions (anions, negatively charged), which are interchangeable.
Annexins	Group of potassium-binding proteins that interact with acidic membrane phospholipids.
Annulus fibrosus	Outer fibrotic layer of the intervertebral disk, with overlapping collagen bundles.
Antibiotics	Substances, largely produced by microorganisms, which inhibit the development of, damage or kill other microorganisms.
Antigen	Protein, foreign to an organism, which triggers the production of antibodies upon its introduction.
Antigen determinant	Site on the surface of an antigen to which an antibody specifically binds.
Antibody	Globular protein produced by certain cells of the immune system in response to the presence of foreign proteins.
Antimycotics	Substances which inhibit the growth of fungi.
Antiserum	Animal blood serum, containing antibodies against one (monospecific) or more (polyspecific) antigens.
Apical	Towards the top, luminal.
Aplasia	Underdevelopment of or complete lack of organ development.
Apocrine secretion	Budding secretion from duct cells.
Apoptosis	Programmed cell death.
Aqua destillata	Distilled water, no longer containing ions.
Aquaporins	Water channel proteins, integral membrane proteins that increase the water permeability of the plasma membrane.
Araldite	Water insoluble, relatively soft artificial substance, used in order to embed samples for electron microscopy.
Arginine	Essential amino acid.
Arteriole	The smallest artery, which through constriction or dilation can regulate blood circulation and which branches into capillaries.

Arthrosis	Wear on the joint.
Ascorbic acid	Vitamin C, strong reducing agent.
Aseptic	Germ-free.
Aspartic acid	Aspartate, amino acid.
Astrocyte	Macroglia cell with radiating cell projections, mostly in the central nervous system.
Asymmetric division	Cell division where, as an example, a stem cell divides into one stem cell and one differentiated cell.
ATP	Adenosine triphosphate, works in cells as energy spender and transformer, due to its high-energy phosphate bonds.
ATPase	Enzyme which splits phosphoric acid from ATP, through which energy is set free.
Atrophy	Tissue deterioration.
Autoclave	Steam pressure sterilizer.
Autocrine secretion	Secretion of a factor that acts on the producing cell.
Autologous	Corresponding, such as in the transplantation of cells from the same organism.
Autolysis	Self-dissolution or digestion.
Axon	Long projection stemming from the nerve cell body for the conducting of an impulse.
Acidophilic	Reacts with the acidic groups of particular dyes.
Acidosis	A decrease in the blood pH to under 7.36.
Acinus	Small, sack structure, surrounded by secretory cells.
Basal ganglion	Corpus striatum, originally at the base of the cerebrum (telencephalon) in vertebrates.
Basal medium	Exactly defined basic medium, containing all vital salts, amino acids and vitamins, but without other supplements.
Basic fibroblast growth factor (bFGF)	Mesodermal and neuro-ectodermal growth factor.
Basolateral	Toward the bottom, serosal.
Basophilic	Reacts with the basic groups of particular dyes.
β-oxidation	Main metabolism pathway of fatty acids, leads to β-keto acids.
Biglycan	Small proteoglycan of the extracellular matrix.
Binocular	Magnifying apparatus for spatial illustration.
Biodegradation	Biological breakdown.
Biocompatibility	Compatibility of medical materials (i.e. for implantation).
Biomedicine	Biological research and development for medical application.
Biopsy	Small tissue sample taken from a living organism.
Biotechnology	Processes used by living cells or enzymes for the transformation or production of substance.
Biotin	Vitamin H, growth factor and coenzyme.
Bioavailability	Effective use for a biological system, systematic availability, percent of a substance which reaches its effective site after application.
Bipolar nerve cells	Nerve cell with one axon and one dendrite.

Blastocyst	Stage of the embryo during the fourth to seventh day of development.
Blastoderm	Early embryonic state.
Blastula stage	Stage of the embryonic development, characterized by a single layer of epithelium and its hollow ball shape.
Blebbing	Characteristic change in the surface membrane during apoptosis (cytosol residues surrounded by largely intact cell membrane).
Blot membrane	Membrane, such as nitrocellulose, used for protein transfer in Western blot assays.
Blood–brain barrier	Partially permeable barrier between the blood and gray matter, mechanism for protecting nerve cells from damaging substances.
Blood coagulation	Enzyme-induced clotting of the blood for wound closing.
Blood platelet	Thrombocyte, blood platelets play a central role in clotting and the closing of damaged blood vessels.
Body segment	Part of the body, disk-shaped anlage during embryogenesis.
B-lymphocyte	Antibody-producing cell of the immune system.
BM-40	Osteonectin, calcium-binding protein in bone.
BMOC (Brinster's modified oocyte culture medium)	Cell culture medium.
BMP	Bone morphogenic protein, growth factor.
Bone marrow	Special, mainly blood cell-building tissue in the pars spongiosa of bone.
Booster effect	Increased effect through repeated treatment.
Bouin solution	Fixing solution for light microscopy preparations.
Bradytrophic	Slower metabolism.
Branching morphogenesis	Development of branched structures at the end of projections.
Bridge molecule	Molecule which connects or conducts electron transfer between molecules.
BSE	"Bovine spongiform encephalopathy", (mad cow) disease caused by infectious proteins (prions).
Buffer	Solution with a stable pH, even when hydrogen or hydroxyl ions are added.
Buffer all	Commercially available buffer mixture.
Buffer media	Buffer that serves as tissue culture media.
Brush border	Line of plasma membrane bulgings on the apical surface that increases the resorptive surface of the cell.
Bypass	Alternative supply path.
Cadherin	Integral membrane protein involved in calcium-mediated cell adhesion.
Calmodulin	Ubiquitous, highly conserved calcium-binding protein.
Calpain	Calcium-activated cytosolic protease.
Calcification	Mineralization, resulting in mechanical hardening of bone.
Canaliculi	Smallest penetration spots, as at the beginning of the gall duct system in the liver.
Channel structure	Penetration of liquids.
Capillary	Smallest blood vessel.
Cardiomyocyte	Adult heart muscle cell.
Carcinogen	Cancer causing.
Carcinoma	Tumor originating in epithelial tissue.

Catalase	Enzyme that catalyzes the reaction of hydrogen peroxide into water and oxygen.
Catalysis	Acceleration process in synthesis.
Cathepsin	Protein-cleaving enzyme.
Calcification	Formation of calcium phosphate deposits in tissue.
Capsula fibrosa	Fibrous caspula in the kidney.
Carbohydrate	Compound with the formula $C_n(H_2O)_n$.
Cardiomyocytes	Heart muscle cells.
Cartilage oligomeric matrix protein	Protein of the thrombospondin family in the cartilage matrix.
Casein	Important protein in milk.
Caspase	Proteases that build the effector machinery in apoptosis.
CEA	Carcinoembryonic antigen, a tumor marker.
Ced gene	Gene involved in apoptosis.
Cell bank	Collection of genetically verified cells that can be recruited as a cell supply.
Cell biology	Study of the molecular processes in cells.
Cell debris	Cell fragments.
Cell differentiation	Specialization of cells in terms of characteristic properties.
Cell	Basic unit of an organism.
Cell nucleus	Cell organelle in which the code for the formation of cellular molecules is localized.
Cell clone	Genetically identical cells originating from the same cell.
Cell contacts	Structures that connect cells mechanically or functionally.
Cell culture, continuous	Cell culture subcultivated more than 70 times.
Cell culture, primary	Freshly isolated cells after the first subcultivation.
Cell line	Cells which can be cultivated due to their permanent ability to proliferate.
Cell organelles	All cell structures with endergonic energy metabolism (i.e. require an input of energy to make chemical bonds) that are enclosed by membranes.
Cell polarization	Property of epithelial cells whereby absorption and release happens on opposite sides of the cell.
Cell stem	Cells from primary or continuous culture that were selected or cloned for specific properties or markers.
Cell suspension	Cell culture where cells are not in contact with the flask surface and therefore float.
Cell therapy	Treatment where cells are used as a therapeutic.
Centrifugation	Separation of suspended particles by centrifugal forces.
Ciliated epithelia	Epithelia, characterized by kinocilia on its apical membrane.
Citric acid cycle	Cycle in the center of metabolism connected to the respiratory chain, responsible for the energy supply, oxidative metabolism of carbohydrates, fats and proteins.
Cerebellar nuclei	Aggregation of specialized cell bodies in the cerebellum.
Clone	Offspring, generated by division of a single mother cell.
Cloning	To raise clones.
Cloning dish	Dish with multiple wells, suitable for cloning.

Cloning cylinder	Special hollow cylinder for isolating single cells.
Collecting tube	Tubular structure, where urine composition can be finally modified.
Cortex	Area where the brain neurons are located.
Cuboidal epithelia	Epithelial cells of the same height and width, cubical shape.
Cytokeratin 19	Intermediate filament expressed in the renal collecting tube epithelium, for example.
Cytokine	Peptides with a signal function made by cells (i.e. growth factors, inflammation mediators).
Cytoplasm	Membrane-free substance of a cell, consisting of water, proteins and numerous ions; individual organelles are embedded in the cytoplasm.
Cytoskeleton	Entirety of the skeletal elements of a cell; the most important parts are actin filaments, microtubuli and intermediate filaments.
Cytotoxic	Poisonous, harmful to the cell.
Chelator	Circular molecule that binds metal ions in a multidental way (e.g. EDTA).
Chemotaxis	Guided movements of cells towards or away from a source, induced by chemical triggers (positive or negative chemotaxis).
Chief cells	Special cell type in the kidney collecting duct.
Chitosan	Polysaccharide from chitin.
Cholesterol	Base substance of steroid hormones and bile acids, is taken in by food (i.e. animal fat) or produced by the body.
Choline chloride	Vitamin B4, important for liver function and fat metabolism.
Chondroblast	Cell that forms cartilage matrix.
Chondroitinase	Chondroitin- and dermatane sulfate-cleaving enzyme.
Chondrone	Chondrocyte and the immediately surrounding matrix.
Chondronectin	Glycoprotein of the cartilage territory.
Chondrocyte	Cartilage cell surrounded by cartilage matrix.
Chorda dorsalis	Early embryonic precursor of the spine.
Chromatin	Fluffy thread-like structure in the nucleus, consists of DNA and specific basic proteins called histones.
Chromosome	Organized form of genomic DNA, limited part of the genetic material in the nucleus of each cell, on which a fixed number of genes are located.
Claudine	Paracellular protein as part of the tight junction.
Co-culture	Culture of various cells or tissue in the same container.
Collagenase	Collagen-cleaving enzyme.
Colon	Large intestine, between ileum and rectum.
Connective tissue	One of the four basic tissues, consists largely of intercellular substance, and fulfills support and metabolic functions.
Connexins	Tunnel-forming proteins in gap junctions.
Connexon	Tunnel formed from connexins in gap junctions.
Coomasie blue	Blue dye that colors proteins.
Cofactor	Participating molecule.
Collagen	Proline-rich protein, main constituent of the mesenchymal extracellular matrix, categorized in fibrillar and non-fibrillar collagens.
Collagenase	Enzyme that metabolizes collagen.

Colloid	Consists of thyroglobulin, abundant in the follicle cells of the thyroid gland.
Colonic crypt	Folding into the lamina propria of the mucosa of the large intestine.
Coma	Condition of deep unconsciousness.
Compartment	Microscopic reaction space within a cell.
Competence	Ability of cells to react to a morphogen within a certain time frame.
Complement	Complex system of the immune system, consisting of at least 17 blood plasma proteins, various activators and inhibitors; the complement supports the action of T- and B-lymphocytes.
Complement factors	Various proteins of the immune system.
Composite materials	Construction of a scaffold, for instance, from a variety of materials.
Confluent growth	Homogenous covering of a surface with tight cell contacts.
Construct	Artificially built tissue.
Contraction	Contraction of cells or tissue.
Copolymers	Polymers formed from more than one type of monomer.
Cornea	Transparent epithelia in the front wall of the eye.
Granule cell	Cell type in the central nervous system.
Cross-contamination	Distribution of a contamination from one culture dish to another.
Cryoprotectivum	Antifreeze.
Cryostat	Machine to prepare thin cryosections for microscopy.
Cryosection	Section prepared by a cryomicrotome.
Crypts	Epithelial invagination in the lamina propria.
Culture container	Microreactor for the preparation of artificial tissue.
Culture medium	Nurturing fluid for cell and tissue culture.
Culture cabinet	Incubator for tissue culture.
Cryo-tube	Tube for freezing samples.
Cyclooxygenase (COX)	Enzyme of arachidonic acid metabolism.
Diacylglycerol (DAG)	Important in signal transduction.
Decorin	Small proteoglycan that binds collagen fibers.
Dedifferentiation	Loss of specialization, regression into a more or less embryonic condition.
Degeneration	Degradation of organs and tissue.
Decubitus	Ulcer or necrosis caused by pressure, "bed sores".
Dendrite	Relatively short, peripherally largely branched cytoplasmic extension of bipolar and multipolar nerve cells; mostly more than one per cell.
Dental enamel	Inorganic material that forms the tooth's surface.
Derivative	Structure which stems from development.
Dermatome	Area of innervation of the skin, by individual spinal ganglia.
Dermis	Skin.
Deaminization	The splitting of an amino group from a molecule.
Desmin	Protein of the cytoskeleton, building block of one group of intermediate filaments.
Desmosome	Mechanical cell–cell contact in which intermediate filaments connect to one another.
Detergent	Tensid, active at the surface.

Determination	Decision during embryogenesis (development).
Dextran	Highly molecular polysaccharide.
Dialysis	Process of cleaning the blood, through the exchange of particles in solution, over a semipermeable membrane (permeable only to molecules of lower molecular weight).
Diastole	Phase of heart muscle relaxation in which the heart is filled with blood.
Differentiation	Development of a cell from the embryonic condition into a specialized cell, with specific form, function and metabolism.
Dictyosome	Cell organelle assembled from stacks of membrane cisterns; altogether, dictyosomes make up the Golgi apparatus.
Dimer	A molecule made up of two separate molecules.
Dimethylsulfoxide (DMSO)	Organic solvent used to protect against crystallization during the freezing process.
Dispase	Metalloproteinase of bacterial origin.
Disse space	Space between endothelia of the liver sinusoid and the liver cell layer.
Diuresis	Increased urine output.
DNA	Deoxyribonucleic acid, carrier of hereditary material.
Domain	Smallest unit of a protein, with a defined and independently folded structure.
Dopamine	Neurotransmitter in the brain, adrenal glands and sympathetic nerve endings.
Double cortin	Microtubule-associated protein expressed in nerve cells.
Drug delivery	Delivery of an active substance to its target.
Dry sterilizer	Air-heated oven.
Duodenum	Part of the small intestine where digestion juices from the liver and the pancreas are channeled in.
Dura	Outer meninges, directly under the skull and partly grown into it.
Dynein	Motor protein that can actively transport (using ATP) molecules or vesicles along microtubules.
Dystrophin	Muscle-specific protein, absence of this protein upsets the calcium balance, which causes failure of the cell.
EGF	Epidermal growth factor.
Eicasonoid	Biologically active substances stemming from arachidonic acid, such as prostaglandins, thromboxanes and leukotrienes.
Embedding resin	Resin into which sample is poured.
Ectodermal	Stemming from the outer germ layer (of the embryo).
Elastin	Main component of elastic fibers, collagen-like protein.
Electron microscopy	Microscopy using of beams of electrons instead of visible light.
Electron transfer chain	Series of electron carriers in the inner membrane of the mitochondria; energy is set free through the transfer of electrons and saved in the form of ATP.
ELISA	Enzyme-linked immunosorbent assay, test in which a bound antibody is recognized by another antibody; the second antibody is marked by the enzyme and becomes visible through the reaction.
Embolus	Material which can leads to the closure of a blood vessel, such as a thrombus or air.

Embryo	Developing child in the womb, before the completion of organ development in the third week of pregnancy.
Embryogenesis	Embryonic development.
Embryonic cell	Unspecialized cell that maintains all or most basic characteristics and which has the potential to differentiate in all directions.
ENaC	Epithelial sodium channel.
Endocrine gland	An organ which secretes hormones into the blood circulation.
Endoplasmic reticulum	Intracellular membrane system in the cell, with transport function.
Endosome	Endosomes develop during the endocytosis of macromolecules; they fuse with primary lysosomes, where macromolecules are broken down hydrolytically.
Endothelial cell	Flat cell type which makes up the simple squamous epithelia lining blood vessels.
Entactin	Protein of the cytoskeleton.
Enterochromaffin cells	Chromaphil and silver-phil cells found dispersed throughout the esophageal and gastrointestinal mucosa, as well as in the gall duct; part of the diffuse endocrine system, building polypeptide hormone, gastrin, secretin and somatostatin.
Enterochromaffin cells	Serotonin-producing cells of the gastrointestinal tract.
Enterocyte	Brush border cell which takes part in intestinal resorption.
Entodermal	Stemming from the inner germ layer of the embryo.
Ephrin	Signaling molecule involved in controlling axon growth.
Epicardium	Outer layer of the heart.
Epicondylus	Bone projection of the condylus for muscle attachment (origin or insertion).
Epidermis	Outer layer of the skin.
Epigenetic	Genetic information found in the cytoplasm rather than the nucleus.
Epiglottis	Structure which covers the glottis (the opening at the upper part of the larynx).
Epilepsy	Cerebral cramps or convulsions.
Epithelia	Tissue type which covers the inner and outer surfaces of organisms in a mosaic pattern, practically without any intracellular matrix between cells.
Epitope	Antigen determinant.
Epon	Particular water insoluble resin.
Equilibration	Adjustment to a condition of balance.
ERK	Extracellular signal-regulated kinase, enzyme of intracellular signal transduction.
Erythropoetin	Growth factor produced in the kidneys that stimulates blood cell production.
Erythrocyte	Red blood cell that does not contain a nucleus, carries out oxygen transport with its hemoglobin.
Estrogen	Hormone.
Exocrine glands	Organs that secrete hormones outwardly, onto the skin or into the intestine.
Exon	Coding or information carrying section of a gene.
Exocytosis	Extrusion of particles from the cell, through fusion of lipid vesicles with the plasma membrane.

Extracorporeal	Outside of the body.
Extraction	Removal of a particular component of a solid or fluid molecular mix, using the appropriate solvent.
Extracellular matrix	Proteins synthesized outwardly by cells that form a mesh or layer on the cell surface.
FCS	Fetal calf serum [also known as fetal bovine serum (FBS)].
Feeder cells	Cells with a "nursing" function that provide sensitive cells with a growth factor-supporting environment.
Fibrin	Fibrous blood protein, builds a fibrous net during clotting.
Fibrinogen	Precursor of fibrin.
Fibroblast	Connective tissue cell involved in the synthesis of intercellular substances.
Fibronectin	Extracellular protein that interacts with various macromolecules, such as collagen, fibrin, heparin and plasma membrane proteins.
Fibrocyte	Resting form of a fibroblast after synthesis activity.
Fibulin	Calcium-binding protein of the ECM.
Filter cartridge	Sterile membrane filter insert for pressure filtration.
FITC	Fluorescein isothiocyanate, amino-reactive fluorescent dye.
Fleece	Fiber material.
Fleck desmosome	Fleck-shaped, highly stabile cell–cell contact.
Fluorescence microscope	Microscope by which the object is excited with a chosen wavelength and the emitted fluorescent light from the object reaches the viewer by a separated pathway.
Fluorochrome	Molecule that emits light when excited with high-energy radiation.
Follicle	Bubble-shaped structure formed out of epithelia.
Fetus	Description of an embryo after completion of organ development.
Freund's adjuvant, complete (CFA)	Used for the increase and modification of the immune response with low antigen concentrations.
Frozen section	Tissue section prepared by a cryomicrotome.
G-protein	Guanine nucleotide binding protein of the cell on which receptors, such as neurotransmitter receptors, are coupled.
Gap junction	Belt-like connection between neighboring cells.
Gastrin	Peptide hormone which stimulates the production of stomach acid and is produced by G-cells of the stomach mucosa.
Gastrulation	The movement of yolk-rich cells, as well as mesoderm and entoderm cells toward the inner embryo during the building of the gastrula.
Gas flow control unit	Apparatus which automatically switches to a full tank when the first one is empty.
Gene switch	Regulates the expression of a protein.
Genital ridge	Thickening of the coelom epithelia on the media side of the urogenital fold.
Genomics	The systematic analysis of the genome.
Gentamycin	Wide-spectrum antibiotic.
Germ layer	Original tissue, consisting of ectoderm, entoderm and mesoderm.
Gonads	Synonym for testis and ovaries.
Gland	Cell or tissue structure which builds a particular substance to be secreted outwardly or inwardly into the blood or lymph system.

Glia cells	Connective tissue cells of the nervous system with support and protective functions.
Globular proteins	Proteins with globular form due to the clustering of amino acid side chains.
Glomerulus	Globular-shaped capillary loop in the kidneys, part of the filtering system.
Glucagon	Hormone produced in the pancreas that causes an increase in the blood sugar level.
Glucocorticoids	Hormones of the adrenal glands, such as hydrocortisone.
Gluconeogenesis	Glucose synthesis pathway starting from non-carbohydrate precursors such as lactate or amino acids.
Glutaraldehyde	Frequently used fixing solution, good for maintaining structures such as the cytoskeleton.
Glycerol	Glycerin, the simplest triple alcohol.
Glycine	Amino acid.
Glycogen	Energy storage made up of carbohydrate chains in the cell.
Glyocalyx	Special structure of glycoproteins and glycolipids on the cell surface.
Glycolipids	Conglomerate of oligosaccharides and lipids in the membrane (ganglioside, sphingomyelin).
Glycoproteins	Proteins with sugar residues.
Glycosylation	Transfer of sugar residues in the biosynthesis of glycoproteins.
Goblet cell	Goblet-shaped, mucin-building epithelial cell (isolated or in groups) of the intraepithelial glands.
Golgi apparatus	Cell organelle made from multiple stacks of flatly pressed membrane sacks (dictyosomes) and vesicles (Golgi vesicles), location of protein modification and mucous production.
Golgi cell	Golgi type I nerve cells have long axons like the motor neurons of the spine or the Purkinje cells of the brain; Golgi type II cells, on the other hand, have short axons and are found as interneurons.
Good Manufacturing Practice	GMP; working according to clearly defined protocols and norms.
Gradient centrifugation	Centrifugation in a linear density gradient in centrifuge tubes; the particles to be separated move to a position corresponding to their density.
Granula	Microscopically small granules that contain synthesized material.
Granulation tissue	Blood vessel-rich new tissue in wound healing.
Grb	Adaptor protein.
Growth factors	Substances that promote growth and proliferation.
Growth hormone	Somatotropin, hormone that promotes growth in length.
GSK	Glycogen synthase kinase.
GTPase	Guanosine 5′-triphosphate (GTP) cleaving enzyme.
Guiding	Experimental directing of cells on a scaffold.
Hematopoiesis	Formation of blood cells.
Hemoglobin	Marker protein of red blood cells – consists of the protein globin and the non-protein heme; heme holds the iron atoms that bind oxygen.
Hemocytometer	Neubauer chamber to determine cell numbers, counting chamber.
Haversian canal	Central canal in the osteon of long bones.
Heat shock protein	HSP, special protein formation in adaptation and under stress.

Hemidesmosome	Contact zone of epithelial cells with the basement membrane.
Hemicyst	Aggregation of epithelial cells without the formation of a continuous basement membrane or a bulb-shaped bulge, called a dome.
Henle's loop	Part of the nephron.
Heparan sulfate	Carbohydrate part of proteoglycans.
Heparin	Heavily sulfated glycosaminoglycan.
Heparinase	Enzyme that cleaves heparin.
Hepatocyte growth factor	Growth factor.
Hepatocyte	Parenchymal cell of the liver.
HEPES	4-(2-Hydroxyethyl)-1-perazine-ethane-sulfonic acid.
Heart valve	Biological valve between the atrium and ventricle of the heart.
Heart tube	Position of the developing heart; during development consisting of two fields that form the endocardium, the myocardium and the pericardium.
Heterodimeric molecules	Compounds consisting of two non-identical subunits.
Heterophilic	Molecules that bind other molecules.
Dorsal root	Area that receives the sensory input from the spine.
Hippocampus	Area of the temporal lobe of the brain.
Histiotypic properties	Tissue-specific properties.
Histoarchitecture	Morphological structure of tissue.
Heat deactivation	Special treatment of sera for the preparation of culture media.
Hollow fibers	Fiber with a lumen.
Holoclone	Special population of cells in stratified epithelia.
Holocrine	Secretion of synthesis products and simultaneous death of the endocrine cell.
Hormone	Specific molecule with information for a cell; a messenger formed by the organism for regulation and coordination of physiological events.
Hormone receptor	Special molecule that recognizes and binds a hormone.
Humoral	Information transfer through a molecule by blood or interstitial fluid.
Hyaline cartilage	Special connective tissue.
Hyaluronidase	Enzyme that digests hyaluronic acid.
Hyaluronic acid	Component of the glycosaminoglycans.
Hybrid	Fusion product of two cells.
Hybridoma	Fusion product of two cells for the production of monoclonal antibodies.
Hybridoma cells	Fusion product of a myeloma cell and a cell that produces antibodies.
Hydration	State of high water content.
Hydrocortisone	Steroid hormone from the adrenal cortex.
Hydrogel	Special ECM made from synthetic material.
Hydrolysate	Protein fragmentation product.
Hydrolysis	Cleavage of a molecule under water consumption, usually by an enzyme.
Hydrophilic	Property of substances with polar groups to from aqueous solutions or bind water; molecules with water-attracting properties.
Hydrophobic	Property of substances without polar groups that from a biphasic system in the presence of water.

Hydrostatic pressure	Pressure produced by a liquid in a resting state.
Hydroxyapatite	Molecule important in the mineralization of bone, dentin and enamel.
Hydroxyapatite crystals	Visible primary structure in the mineralization of bone, dentin and enamel.
Hydroxy group	Hydrophilic group in a molecule.
Hyperplasia	Proliferation of living matter by increase in cell number.
Hypertrophy	Proliferation of living matter by increase in cell volume.
Hypoblast	Embryonic stage before germ layer development.
Hypophysis	Pituitary gland, central for the regulation of organs and tissue by hormone secretion.
Hypoxanthine	purine base, intermediate in the nucleic acid metabolism.
Hypoxanthine guanine phosphoribosyl transferase	Enzyme in the DNA biosynthesis pathway.
Hypoxia	Insufficient supply with oxygen.
ICAM	Intracellular adhesion molecule.
IGF1R	Insulin-like growth factor receptor.
Ileum	Part of the small intestine.
Immediate-early growth response gene	Gene response within minutes after stimulating the cell.
Immunoglobulin	Globular protein that acts as an antibody and binds xenobiotics.
Immunohistochemical	Investigation of cells and tissue by immunochemical methods, usually antibodies.
Immune complex	Aggregation of an antibody with an antigen.
Immunofluorescence test	Analysis of an antigen (or first antibody) with an antibody (or second antibody) tagged with a fluorophore.
Immunology	Science regarding the immunologic defense mechanisms of the body.
Immunosuppression	Therapy to avoid the rejection of transplanted organs or tissue.
Immune suppressant	Medication that which suppresses the immune system of the body.
Implant	Material that supports regeneration procedures.
Implantation	Insertion of an implant.
In situ hybridization	Specific analysis of DNA and RNA in a histological sample by using short DNA/RNA probes.
In vitro	"In glass", e.g. performed in the laboratory.
In vitro fertilization	Fertilization in a cell culture dish.
Induction	Induction of a growth or differentiation process in a cell or group of cells.
Induction stimulus	Trigger of induction with morphogenic compounds.
Industrial cell culture	Cell culture and preparation of synthetic products on an industrial scale.
Incubator	Container to provide a constant environment for cultures.
Inner ear	Part of the hearing organ.
Innervation	Connection of a nerve cell with identical or different tissue.
Inositol	Vitamin, attributed to the vitamin B2 complex.
Insect cell culture	Culture with insect cells, often used to prepare recombinant proteins.
INSR	Insulin receptor.
Insulin	Hormone of the carbohydrate metabolism.

Integrin	Surface molecule abundant in many cell types, for adhesion, interaction, and signal transduction.
Intercalated disks	Area in which neighboring heart muscle cells connect mechanically and where the cytoplasm communicates.
Interleukins	Cellular hormones, mediators of the immune system.
Intermediate filaments	Protein filaments of the cytoskeleton (8–10 nm diameter), that are wider than actin but thinner than microtubuli.
Intermediate zone	Central cell zone in the stratified epithelia.
Internist	Specialist in internal medicine.
Interphase	Phase in the cell cycle between two cells divisions.
Interstitial lamella	Residual general lamella of the Haversian system filling bone gaps.
Interstitium	Space between tissue and cells filled with liquid.
Intercellular space	Space in between cells.
Intercellular substance	As observed by light microscopy, a structureless mass with embedded fibrous proteins in between cells.
Intraperitoneal	Within the peritoneum.
Intravenous	In the veins.
Intravital	Alive.
Intron	Non-coding in between coding sequences of genes.
Inverted microscope	Microscope in which the light beam of a classical microscope has been turned around – the objective is brought to the stage from below.
***In vitro* experiment**	Experiment under culture conditions without an animal.
***In vivo* experiment**	Experiment with a human or animal specimen.
Involucrin	Protein in the differentiation of keratinocytes.
Isoelectric focusing	Method for separating amphoteric compounds depending on their iso-electric point.
Isoelectric point	IP, typical value for the charge properties of proteins in an ampholyte gradient; pH value at which amphoteric compounds appear electrically neutral, due to equally strong dissociation of their acidic and basic groups.
Isoform	Of the same kind.
Iso-osmotic	Cellular environment under natural conditions.
Isotonic	With the same osmotic pressure.
Ito cell	Special cell type of the liver within Disse space.
JAM	Junctional adhesion protein, occurs in tight junctions.
JE gene	Induced by PDGF; also called MCP-1.
Jejunum	Part of the small intestine.
JNK	c-Jun N-terminal kinase.
Joint	Mobile connection.
Jun	c-Jun N-terminal kinase, see JNK.
Keratin	Protein of the cytoskeleton.
Keratinocytes	Keratin-producing skin cells.
Keratohyaline granula	Particles that contribute to keratinization.
Keratoplasty	Renewal of the cornea in the eye.

Keratoprosthetic	Replacement of the cornea in the eye.
Keratocytes	Epithelial cells of the cornea.
Ki67 protein	Also called MIP1, expressed during mitosis, not expressed during interphase.
Kinase	Phosphorylating enzyme with function in signal transduction.
Kinesin	Protein with transport function.
Kinocilia	Cell organelle to provide mobility, occurring in some epithelia.
Knock-out animal	Animal with an experimentally induced missing gene function.
Lacuna	Residence of chondrocytes.
L1	Adhesion molecule, expressed by many axons.
Lactate	Metabolic product.
Lamellar bone	Special bone type.
Laminar air flow	Tissue culture hood for sterile procedures.
Laminin	Glycoprotein of the basal lamina.
Langerhans cells	Hormone-producing cells of the pancreas.
Laser scanning microscope	Special microscope for the recognition of three-dimensional structures.
Lateral	To the side, away from the center.
Lectin	Proteins reacting specifically with certain carbohydrates and glycoproteins of plant origin.
Lentigo senilis	Age spots.
Leukotriene	Intermediate in the metabolism of arachidonic acid.
Leukocytes	"White blood cells", classified as granulocytes, lymphocytes and monocytes.
L-Glutamine	Amino acid with a key position in the amino acid metabolism.
Light microscopy	Projection of histological structures.
LIF	Leukemia-inhibiting factor.
Ligand	Binding molecule.
Limbus	Interface between cornea and sclera, area with stem cells in the eye.
Lipase	Fat-cleaving enzyme.
Lipofuscin granula	Pigment granula in the cell, mostly metabolic waste products, compartmentalized in lysosomes.
Lipoproteins	Protein with a lipid moiety.
Logarithmic growth phase	Growth phase, where the cell number increases 10 times per unit time.
Lumen	Cavity of a hollow organ, apical boundary of epithelia.
Lumican	Corneal keratan sulfate proteoglycan.
Lymphocyte	Cell of the immune system.
Lysine	Amino acid, often glycosylated in collagen.
Lysosomes	Vesicular organelles with a specific set of enzymes for intracellular digestion.
Lysozyme	Bactericidal enzyme from Paneth cells.
MRNA	Messenger RNA, information carrier for the production of proteins in cells.
Mad cow disease	BSE-like disease.
Macromolecule	Highly polymeric molecule made from more than 1000 atoms.

Macrophages	Special cell type of the blood; long-lived giant cells derived from monocytes, which can phagocytose foreign substances.
Macroscopy	Anatomical lesions at the corps.
MALDI-TOF	Matrix-assisted laser desorption and ionization time of flight mass spectrometry.
MAPK	Mitogen-activated protein kinase.
Marker protein	Characteristic protein useful for the identification of cell differentiation.
Myelin sheath	Myelin cover of nerve fibers.
Mass spectrometry	Method to detect molecules.
Mast cells	Special blood cell type.
Matrigel	Extracellular matrix synthesized by tumor cells.
Matrix	ECM, extracellular matrix.
Matricellular proteins	Special proteins between the plasma membrane and the ECM.
MCAF	Monocyte chemotactic activating factor.
MCP-1	Monocyte chemo attractant protein-1.
M-CSF	Macrophage colony stimulating factor.
MDCK cells	(Madin-Darby canine kidney) cell line, established from the kidney of a cocker spaniel bitch, isolated by S. H. Madin and N. B. Darby in 1958; cannot be assigned to one particular cell type of the nephron due to mixed characteristics.
Mechanoreceptor	Sensor for mechanical and elastic deformation.
Media design	Development of novel culture media.
Medulla	Marrow, inner part of an organ.
Medulla	Area of the brain with ascending and descending fibers.
Medulloblastoma	Fast growing, non-differentiated tumor in the cerebellum.
MEF	Murine embryo fibroblast.
Melanocyte	Special cell type of the skin.
Meltrine	Member of the ADAM metalloproteinases in the ECM.
Membrane depolarization	Alteration of the electrical properties of the plasma membrane.
Memory cells	Cells of the immune system, responsible for the secondary immune response to repeat infections.
Membranous bones	Desmal bones of the cranium, including facial bones.
Merocrine	Special kind of secretion from endocrine cells.
Mesangia	Supporting tissue in the glomerula.
Mesangial cell	Special cell type in the glomerular mesangia.
Mesenchyme	Embryonic connective tissue, most originate from the mesoderm.
Mesoderm	One of the three germ layers of the embryo.
Mesothelia	Striated epithelia in the chest and abdomen.
Messenger RNA (mRNA)	Translatable copy of genes, transported from the nucleus to ribosomes in the cytoplasm, where translation into protein happens.
Mest	Mesoderm-specific transcript, formerly Peg1.
Metabolic engineering	Experimental optimization of metabolism under culture conditions.
Metabolite	Low-molecular-weight substance, occurring in biological metabolism, often intermediate or final steps.

Metalloproteinases	Family of enzymes involved in ECM turnover.
Metastasis	Tumor cells released from a tumor.
Methylation	Transfer of methyl groups to molecules.
Migration	Spontaneous change in (cell) location.
Microfibril	Collagen fibril, detectable by electron microscopy.
Microfilament	Filament of the cytoskeleton made from actin.
Microglia	Small, long cells from the grey and white matter of the central nervous system.
Microreactor	Apparatus for the production of artificial tissue.
Microscopy	Visualization of structures.
Microstructures	Scaffolds with particular surface properties.
Microtubules	Tubular cytoskeletal structures consisting of 13 protofilaments, which are in turn made from tubulin dimers.
Microvilli	Special surface differentiation; finger-shaped, mostly non-branched bulges of the plasma membrane.
Milieu	Environment for cells and tissue.
Mitochondria	Special cell organelle for energy generation.
Mitogen	Substance that induces cell proliferation.
Mitosis	Cell division.
Mnk1	MAP kinase interacting kinase 1.
Modulation	Alteration of properties.
Molecular weight	Size of a molecule.
Monoclonal	Derived from a single clone.
Monolayer	Single sheet of cells growing on a surface.
Monocyte	Special cell type in the blood.
Morphogen	Substance that influences the development of form.
Morphology	Study of the structure and form of living beings and their organs.
Morula stage	Early stage in embryo development.
Motif	Defined amino acid sequence of peptides and proteins.
Motor neuron	Neuron that transmits motor impulses.
Motor endplate	Synaptic connection between the axon of the motor neuron and a muscle fiber.
Motor protein	Intracellular protein involved in transport and movement.
MSOS	Mammalian son of sevenless, guanidine nucleotide exchange factor for Ras and Rac.
MTL	Methane thaniol.
Mucosa	Epithelia that covers inner cavities of the body.
Mucosal	Mucus containing.
Multiple sclerosis	Fairly frequent disease of the central nervous system.
Multipolar neuron	Neuron with many dendrites and one axon.
Multivacuolar	Cells with numerous vacuoles.
Muscular dystrophy	Degradation of muscle tissue.
Muscle fascia	Connective tissue surrounding muscles.

Muscle contraction	Shortening of the muscle.
Mutant	Cell with an altered set of chromosomes.
Myelin	Membrane lipids of a nerve fiber forming the myelin sheath.
Myelin sheath	Cover of an axon.
Myeloma cells	B-lymphocytes, tumor cells.
Myf5	Myogenic regulatory factor 5.
Mycoplasma	Wall-less prokaryotes, that reside as parasites in eukaryotic cells; frequent contamination in animal cell cultures, identified by electron microscopy, histochemical and immunological methods.
Myoblast	Immature muscle cell.
MyoD	Myogenic transcription factor.
Myofibril	Contractile element.
Myofibroblast	Contractile fibroblast.
Myogenin	Myogenic transcription factor.
Myocardium	Heart muscle.
Myosin	Second major component of the actin–myosin system, marker protein of muscle cells.
Myosin head	Molecular part of the contractile apparatus.
Myotome	Muscle segment in embryo development.
Na/K-ATPase	Active sodium/potassium pump.
Metanephros	Remaining kidney organ established in the 4th week of development.
NCAM	Neural-cellular adhesion molecule.
Nck	Adaptor protein.
Necrosis	Induced cell death (*cf.* Apoptosis).
Neocortex	Part of the cortex.
Neoplasia	Formation of new tissue, tumor formation.
Nephrotome	Location of kidney development during embryonic development.
Nerve fiber	Dendrites and axons.
Netrins	Chemotactic cellular protein.
Neubauer counting chamber	Hemocytometer.
Nuclear fluorescent staining	Coloring with fluorophores, which locate to the nucleus (i.e. DAPI).
Neural	Affecting the nerve system or its function.
Neural plate	Structure during development.
Neural tube	Structure of the central and peripheral nerve system during development.
Neuroblastoma	Tumor of the nerve system.
Neurofilaments	Group of intermediate filaments of the cytoskeleton.
Neuroglia	All cells of the nervous system except neurons.
Neurology	Study of neural diseases.
Neuron	Nerve cell.
Neuropilin	Co-receptor for semaphorins.
Neurotransmitter	Messenger substance within synapses.

Neurotrophin	For instance hippocampus-derived neurotrophic factor.
Neurula state	Developmental state.
Nexins	Proteins of the cytoskeleton.
Nicotinamide	Important co-enzyme.
Nidogene	Connecting protein between cytoskeleton and ECM.
NIH	National Institute of Health, Bethesda, MD.
Nissl staining	Special histological staining of neurons.
Nitrocellulose	Nitrated cotton; used as scaffold for epithelia, for instance.
Nitrogen monoxide	Vasodilating molecule.
Node of Ranvier	Myelin-free area between two glia cells.
NrCam	Neural cell adhesion molecule.
Nucleus pulposus	Core part of the vertebral disc.
Nucleic acids	Genetic information carriers in chromosomes and RNA.
Nucleolus	Nuclear body observed in interphase of the cell cycle.
Nutritional	Concerning food or nutrition.
Occludin	Tight junction protein.
Oct4	Transcription factor.
Oligodendrocyte	Myelin-producing cell in the central nervous system.
Oligosaccharide	Carbohydrate consisting of 3–12 monosaccharides.
Oncotic pressure	Osmotic pressure in a colloidal solution.
Ontogenesis	All form-shaping processes from oocyte to adult organism.
Organ	Functional unit consisting of parenchyma and stroma; segregated part of an organism with characteristic location, shape, and function, usually constituted from several tissues.
Organ anlage	First appearance of organ formation.
Organoid	An organ-like structure.
Orthopedics	Medical area of the motility apparatus.
Osmium	Fixation agent for electron microscopy with good structure maintenance, in particular of membranes.
Osmium contrasting	Contrasting of tissue with osmium.
Osmolarity	Measuring unit for dissolved molecules in a solution.
Osmolyte	Osmotically active substance.
Osteoblast	Cell of mesenchymal origin that secrets bone substance.
Osteocalcin	Bone gla protein (BGP), important protein for bone formation.
Osteocyte	Mature osteoblast after enclosure into intercellular matrix.
Osteoinduction	Stimulation of osteoblasts form bone in a scaffold.
Osteoclast	Bone-reabsorbing cells, part of the bone-forming system.
Osteoconduction	Guiding of osteoblasts in a scaffold in a particular direction for bone formation.
Osteon	Smallest functional unit of lamellar bone.
Osteopontin	Important protein in bone formation.
Osteoporosis	Disease of the skeleton, recognizable by a loss in bone density.

Ouchterlony test	Test to determine the immunoglobulin class of an antibody – the unknown antibody is brought into contact with antibodies of various immunoglobulin classes and recognition reactions are monitored.
Oval cells	Stem cells of the liver.
Oxidase	Enzyme that catalyses oxidation or reduction reactions.
Oxygenation	Supply with oxygen.
P130CAS	Adapter protein involved in migration and adhesion of cells.
p53	Special phosphorylated protein.
Paneth cell	Epithelial cell with strongly oxyphilic granules located in the ileum crypts.
Paracrine	In close proximity.
Parathormone	Parathyroid hormone.
Parathyroid gland	Hormone-secreting gland.
Paraxial	Parallel to an axis.
Parenchyma	Functional tissue of an organ.
Parkinson disease	Degeneration of the substantia resulting in reduced levels of the neurotransmitter dopamine.
Partial pressure	Unit for dissolved gases.
Passaging	Subcultivate.
Pasteur pipette	Serves to transfer cells.
Pathogenic	Disease-inducing.
Pax2	Nuclear transcription factor involved in differentiation.
Paxillin	Adhesion molecule that links actin filaments to the plasma membrane.
PBS	Phosphate-buffered solution.
PDGF	Platelet-derived growth factor.
Pellet	Solid formed after centrifugation of a suspension.
Penicillin G	Antibiotic.
Peptide	Polymer made from less than 30 amino acids.
Perfusion culture	Cell culture with continuous supply of fresh culture medium.
Pericardium	"Bag" made from connective tissue covering the heart.
Perikaryon	Body of a nerve cell containing almost all organelles.
Perimysium	Connective tissue layer covering muscle fibers.
Periostium	Connective tissue around the bone, rich in blood vessels and nerve fibers.
Periphery	Area away from the center.
Peristaltic	Rhythmic contraction waves of hollow organs.
Peritoneal cells	Cells of the peritoneum.
Peritoneum	Serous membrane covering the abdominal cavity and abdominal organs.
Perlecan	Proteoglycan of the basal membrane, consisting of five pearl string-like globular sections, which are connected to integrin receptors via heparin sulfate side chains.
Permeability	Property that permits transport of substances, e.g. through a membrane.
Peroxisomes	Vesicular cell organelle with the typical enzymes peroxidase and catalase.
pH	A term describing the concentration of protons.
Phagocytosis	Uptake of solid particles by a cell.

Phenotype	Typical appearance of an organism; the appearance at a certain time point in development.
Pharmacology	Study of medication.
Phase contrast microscopy	Special form of light microscopy, where differences in optical refraction are translated into different brightness levels.
Phenol red	Indicator dye that crudely shows the pH.
Phosphatase	Hydrolase that cleaves phosphate groups, thereby activating or inactivating cellular proteins.
Phospholipid	Molecule with hydrophobic and hydrophilic parts, consisting of a central molecule, fatty acids and a phosphorylated alcohol; building blocks of all biological membranes.
Phosphorylation	Esterification of ortho- or pyrophosphoric acid with OH groups containing organic compounds; activation of cellular proteins by adding a phosphate ester.
Phylogeny	Evolutional development of an organism; the physiological passing through appearances of older stems of an organism in development.
Physiological	Natural, as a natural procedure.
Physiological salt solution	Sodium chloride solution of 0.9% isotonic to blood serum.
Pigment	Colored molecule in the body.
Plasma membrane	Membrane enclosing a cell, consisting of a double layer of phospholipids and other lipids with numerous proteins engulfed.
Plasma cell	Differentiated form of B-lymphocytes as producers of antibodies.
Plasmin	Active protease that cleaves fibrin and basal lamina proteins.
Plasminogen	Inactive precursor of plasmin.
Plasticity	Ability of a cell to acquire additional or different functions.
Placenta	Tissue that supplies the embryo.
Pleura	Serous membrane covering the lungs.
Pluripotence	Property of stem cells to develop into different cell types.
PNA	Glycoprotein from plants.
Podocyte	Cells covering the capillaries of the renal glomeruli.
Polio	Virus disease, predominantly occurring during childhood.
Polycarbonate	Heat-resistant, clear thermoplastic belonging to the technical plastics.
Polyethylene terephthalate (PET)	Aromatic polyester; used to make artificial blood vessels, for instance.
Polyclonal	Derived from several clones.
Polylactide	Biodegradable polymer made from lactate monomers.
Polyglycolide	Biodegradable polymer made from glycolic acid monomers.
Polymer	Macromolecule constructed from uniform monomeric molecules.
Polymorphic	Of various shape.
Polypeptide	Linear polypeptides, connected by peptide bonds.
Polyribosome	Aggregate of several ribosomes that translate mRNA into protein.
Polysome	Polyribosome.
Postmitotic	After proliferation is finished.
Pre-cordal plate	Cranial bulge of the entoderm.

Precipitate	Substance in solid form from a solution.
Pre-pro form	Precursor form of a protein.
Primary immune response	Early reaction to the intrusion of an antigen; B-lymphocytes make first contact with the antigen, form clones and secret specific antibodies.
Primaria dishes	Special tissue culture dishes with positively charged residues on the surface that mimic proteins in order to foster cell adhesion.
Primary culture	Culture of original cells or tissue from an organism.
Primitive knot	Cranial end of the primitive streak.
Primitive streak	Medial bulge of the germinal disk.
Prions	Infectious proteins that induce degenerative brain diseases.
Pro-erythroblasts	Earliest stage in the differentiation of erythrocytes.
Profiling	Determination of cellular properties.
Progenitor cell	Unipotent or bipotent stem cell.
Pro-collagen	Collagen precursor.
Prolactin	Hormone that stimulates the milk production of mammary glands.
Proliferation	Growth by cell division.
Proline	Hydrophobic amino acid.
Propidium iodide	Dye to stain DNA.
Prostaglandins	Tissue hormones made from arachidonic acid; function in pain, fever, inflammation, etc.
Protease inhibitor	Inhibitor of proteolytic enzymes.
Proteases	All enzymes that catalyze the cleavage of proteins.
Proteins	Molecules made from more than 50 amino acids, which are responsible for most biological functions.
Protein biosynthesis	Synthesis of proteins inside cells.
Proteoglycan	Protein containing covalently bound amino sugar chains, part of the ECM, glycoproteins with a core protein, coupled to glycosamine glycan.
Proteolysis	Protein metabolism in the course of physiological protein digestion; a biochemical method.
Proton pump	ATPases that transfer H^+ ions through membranes.
Protrusion	Bulging out.
Pseudounipolar nerve cell	Neuron in the spinal ganglia with a cell extension that splits in T-form shortly after leaving the cell body.
Purkinje cell	Neuron of the cerebellar cortex with a characteristic dendrite tree.
Radial	Beam-shaped.
Radioactive	Radiation releasing.
Radioimmunoassay (RIA)	Radioimmunological antigen assay for the quantitative determination of small amounts of compound.
Raf	Molecule that takes part in signal transduction.
Regeneration	Regeneration of a biological function.
Recombinant	Produced by transformation through genetic technology.
Recombinant protein	Protein produced by transformation through genetic technology.
Relay model	Induction cascade triggered by a signal that results in another signal induction in differentiated neighbor cells.

Renin	Key enzyme of blood pressure regulation formed in the kidney.
Repeats	Reoccurring amino acid sequence.
Repulsion	Rejecting response of a signaling molecule on cell extension growth.
Residual body	Final storage of metabolic products, mostly former lysosomes.
Resorption	Take-up.
RET	Receptor tyrosine kinase.
Reticular fibers	Silver-stainable collagen fibers.
Reticulin	Protein of connective tissue forming reticular fibers.
Reticulocytes	Precursor of erythrocytes that still have remains of the protein biosynthesis apparatus.
Retina	Layers of cells on the posterior part of the eyeball which receive the image.
Receptor	Molecule that induces a cellular response upon binding of a ligand.
Receptor protein	Protein that receives certain signals.
RGD motif	Smallest structural element in the form of a tripeptide that leads to binding of an integrin molecule to collagen.
Rheology	The study of the deformation and flow of matter.
Rho	GTP-binding protein.
Riboflavin	Vitamin B_2.
RNA	Ribonucleic acid, cellular information carrier build from nucleotide building blocks.
Roller bottles	Culture bottles that roll during cultivation, thereby improving gas and nutrition exchange compared to stationary cultures.
S-phase	Replication phase in the cell cycle where DNA synthesis happens.
Saltatory conduction	Fast spreading excitation via a myelinated axon.
Sarcolemma	Plasma membrane of the muscle cell.
Sarcomere	Contractile units of a myofibril.
Satellite cell	Nurturing cells of pseudounipolar neurons.
Scaffold	Three-dimensional carrier material in tissue engineering.
Scaffolding	Cell biological influence on cells on or within an ECM.
Scar tissue	Regenerated tissue.
Schwann cell	Peripheral glia cell that forms myelin sheets around axons.
Scanning electron microscope	Electron-based surface imaging.
Striation	Morphological property of skeletal muscles due to the sarcomere orientation.
Sweat gland	Exocrine gland of the skin that regulates fluid loss over the skin.
Screening kit	Commercial test system to check for the presence of a certain marker (e.g. mycoplasma, production of antibodies, etc.).
Secretion granula	Vesicle that contains products to be secreted.
Secretion product	Product of a gland.
Secondary immune response	Reaction against the repeated intrusion of an antigen, against which a specific antibody has been formed before; there are B-lymphocytes that have synthesized the specific antibody before and which now may react faster and more intensely.

Selectins	Cell–cell adhesion proteins that interact via sugar molecules.
Selection media	Media used to select for a cell clone.
Semaphorins	Class of molecules that guide the growth of axons.
Semi-quantitative	Estimate of the amount based on an incomplete data set.
Sequence analysis	Determination of the amino acid sequence of a protein.
Serous	Terminal parts of glands that produce a watery secretion.
Sertoli cells	Scaffold cell of the gonad epithelia.
Serum	Non-agglutinating, cell-free part of the blood.
Serum batch	Serum package, ready for experimental use.
SHH	Differentiation factor of the hedgehog family.
Sialoprotein	Glycoprotein of the bone matrix.
Sic1	Cell cycle protein.
Signet ring form	Typical cell shape of univacuolar adipocytes in paraffin sections.
Siemens	Unit of electric conductivity.
Signaling cascade	Receptor-mediated intracellular reaction sequence.
Signaling sequence	Amino acid sequence that serves as a biological signal.
Simple squamous epithelia	Epithelia with flattened cells.
Sensory cell	Cell that picks up sensory stimulus.
Slice culture	Culture of thin tissue slices.
Small intestine submucosa (SIS)	Biomatrix.
Somatostatin	A tetradecapeptide of the hypothalamus that inhibits the secretion of somatotropin from the pituitary gland.
Soma	Temporal segments of the paraxial mesoderm that lead to segmental arrangement of the mesoderm.
Sonic hedgehog	Protein that governs limb development.
Sox	Group of 30 transcription factors which are also involved in development.
SPARC	Osteonectin, involved in bone formation.
Salivary gland	Exocrine gland, the products form the saliva.
Spermatogonium	Precursor of sperm cells, occurring in the basal part of the gonad epithelia.
Spinal ganglia	Group of pseudounipolar neurons, which conduct afferent impulses to the central nervous system.
Splicing	Excision of introns from the primary transcript and ligation of the remaining exons.
Spreading	Migration of cells.
Stem cell	Self-renewing cell, its division leads to a cell with complete and/or one with limited developmental potential; cell with the ability to reproduce itself in an unlimited fashion by cell division and with the ability to produce cells with different specializations.
Stem cell niche	Localization where a stem cell resides.
Starch	Primary storage carbohydrate of plant cells which is exclusively built from glucose.
STAT	Transcription factor.
Stem cell factor	Growth factor.

Sterile	Without microorganisms.
Sterile filtration	Sterility generated through micropores.
Strands	Tight junction strands.
Stratum corneum	Uppermost layer of the skin.
Stratum ganglionare	Purkinje cell layer in the cerebellar cortex.
Stratum granulosum	Cell layer of the skin that shows keratohyaline granules.
Stratum moleculare	Outer layer of the cerebellar cortex.
Streptomycin	Antibiotic.
Stria vascularis	Epithelia of the inner ear, which is vascularized.
Stroma	Scaffold tissue of an organ made from connective tissue.
Stromelysin	Matrix metalloproteinase.
Subcultivation	Transferring cells from one flask to the next.
Submucosa	Connective tissue underneath the mucosa.
Subpopulation	Part of a population that generates offspring, but is not connected to the rest of the population.
Substantia	Amorphic component of the intercellular space in connective tissue.
Sucrose	Saccharose, cane sugar.
Superinfection	One infection superimposing another.
Support	Carrier material.
Suspension	Mixture of non-dissolved particles in a liquid.
Suspension culture	Cell culture where cells have no contact to the support or flask.
Synapse	Transmitter-mediated functional connection of two nerve cells.
Syndecan	Integral membrane proteoglycan.
Synovia	Liquid of the joints that feeds cartilage cells among other things.
Syncytium	Polynuclear cells formed by cell fusion.
Systole	Blood release phase of the heart.
Sebaceous gland	Exocrine gland of the skin with holocrine secretion of sebum.
Talin	Actin-binding protein.
Target selection	Selection of the target for axon growth.
Tcf	Transcription factor.
Telomere	Terminal region of a eukaryotic chromosome, which is constantly replicated; telomeres prevent chromosome shortage during replication.
Tenascin	Extracellular glycoprotein with six branches, occurring during development and in tendons.
Tendon	Connective tissue-like connection between muscle and bone.
Tendon cells	Fibrocytes in the tendon, with thin extensions adjusted to the shaped of the fibrous bundle.
Tensin	Actin-associated protein that can be phosphorylated.
Teratocarcinoma	Undifferentiated teratoma, predominantly of the testis.
Terminal differentiation	Generation of specific cell functions after the final phase of development.
Territory	Cartilage cell and surrounding area.
Thermanox	Polymer.
Thrombin	Enzyme that cleaves fibrinogen to fibrin.

Thrombospondin	Extracellular glycoprotein.
Thymidine	Building block of DNA.
Thymine kinase	Enzyme that catalyses the phosphorylation of thymine.
Thymocytes	Cell of the thymus.
Thyroidea	Thyroid gland.
Tight junction	Belt-shaped cell–cell contact between epithelia cells, controls diffusion through the intercellular space.
Time window	Time period during which certain molecules are active during development.
Tissue	Grouping of similar differentiated cells.
Tissue engineering	Production of artificial tissue.
Tissue factory	Modular system in tissue engineering.
Tissue hormone	Hormones produced in individual cells, localized in particular organs or organ systems, rather than in glands.
Tonsil	Organ with immunological function.
Totipotency	Cell that is able to develop into all three germ layers.
Transdifferentiation	Differentiation of one cell type into another.
Transduction	Transmission of a signal.
Transferrin	Often used culture reagent.
Transfilter experiment	Culture method in which cells are supplied from the basal side after reaching confluency.
Transformation	Alteration of genetic properties by incorporating external DNA strands into the host genome.
Transforming growth factor β	Differentiation factor.
Transgene	Higher organisms that carry foreign genes.
Transcription	Re-writing the genetic code from DNA by synthesis into a complementary mRNA sequence during protein biosynthesis.
Transcription factor	Protein that regulates transcription.
Translation	"Translating" genetic information stored in mRNA into the amino acid sequence of a gene-specific polypeptide during protein biosynthesis.
Transmembrane protein	Protein that spans the entire plasma membrane.
Transmitter	Molecules that are released into the synaptic cleft during signal transduction.
Transport protein	Membrane-bound protein that transports certain substances through the membrane.
Triacylglycerol	Form of fat storage.
Tricalcium phosphate	Inorganic material of bones.
Triple helix	Three-dimensional structure.
Trophoblast	Peripherally oriented blastomeres that form the placenta.
Trypan blue	Dye that enters dead, but not living, cells.
Trypsin	Proteolytic enzyme, serine protease.
T-tubulus	Fold of the plasma membrane in muscle fibers.
Tubulin	Structural protein of microtubuli.
Tubulus cell	Epithelial cell of tubular ducts.

Tumor	Increase of tissue volume by pathological cell proliferation.
TUNEL	(terminal deoxyribonucleotidyl transferase-mediated dUTP nick-end labeling) principle; Proof of apoptosis.
Tween 80	Detergent.
Tyrosine kinase	Enzyme that phosphorylates the amino acid tyrosine.
Transitional epithelium	Mucosa of the urinary tract.
Ubiquitin	Highly conserved protein which, when attached to other proteins, induces their degradation.
Ulcus cruris	Ulcer of the legs induced by vascular diseases.
Ultrathin cut	Slice of a thickness between 0.03 and 0.1 μm.
Ultramicrotome	Machine for the preparation of histological slices thinner than 1 μm.
Untouched layer	Location with little mixing.
Univacuolar	Formation of only one vacuole.
Uroplakin	Protein formed in the urothelium.
Urothelium	Epithelium of the urinal tract.
Vaccine	Medication to avoid infections.
Vacuole	Bubble-shaped organelle in the cytoplasm.
Vacuum filtration	Filtration method for small volumes: an aspirator generates reduced pressure that forces the liquid through a sterile filter.
Vascular cellular adhesion molecule (VCAM)	Cell adhesion molecule in vessels.
Vascular endothelial growth factor (VEGF)	Angiogenetic growth factor.
Vascularization	Supply with blood vessels, formation of new blood vessels.
Vascular genesis	Formation of new blood vessels.
Vasopressin	Peptide hormone that regulates the water release of the kidney.
Vectorial transport	Directed transport.
Versican	Extracellular protein formed by fibroblasts.
Vesicle	Small membrane-enclosed cell compartment.
Vimentin	Cytoskeleton protein, building block of a group of intermediate filaments.
Vinculin	Anchor protein of the cytoskeleton.
Viscosity	Flow property.
Visceral	Regarding the organs.
Vital dye	Dye that can enters living cells.
Vitality test	Test of survival.
Vitamins	Essential organic compounds partly synthesized by the organisms partly supplied with the food.
Vitiligo	White, pigment-free skin areas.
Vitronectin	Adhesion molecule occurring on cell surfaces as well as in blood plasma.
Volkmann's canal	Channel through which vessels enter the bone.
Von Willebrand factor	Platelet aggregating factor formed by endothelial cells.
Ventral horn	Motoric fibers leaving the spinal cord.

Western blot	Analysis of antigens by electrophoresis separation, transfer of the separated proteins onto an inert support and recognition by specifically marked antibodies.
Wnt protein	Differentiation factor.
Wolffian body	Mesonephros; transiently formed kidney in embryogenesis.
Wound healing	Physiological closure of a wound.
Xenotransplant	Organs or tissue from donors of a different species.
ZO-1	A tight junction protein.
Zona occludens	Tight junction.
Z stripes	Border between two sarcomeres.
Zygote	Fertilized oocyte.

Companies

ADVANCED SCIENTIFICS, INC.
163 Research Lane
Millersburg, PA 17061, USA
(717) 692-2104
(717) 692-2197
www.advancedscientifics.com

ALLCELLS, LLC
2500 Milvia Street, Ste. 214
Berkeley, CA 94704, USA
(510) 548-8908
(510) 548-8327
www.allcells.com

AMERICAN TYPE CULTURE
COLLECTION
10801 University Blvd.
Manassas, VA 20110-2209, USA
(703) 365-2700
(703) 365-2701
www.atcc.org

AMRESCO INC.
30175 Solon Industrial Pkwy.
Solon, Ohio 44139, USA
(800) 829-2802
(440) 349-1182
www.amresco-inc.com

AMS BIOTECHNOLOGY (EUROPE) LTD.
Centro Nord Sud
Stabile 2 Entrata E
Bioggio, Ticino Switzerland 6934
+41 91 604 5522
+41 91 605 1785
www.immunok.com

APPLIKON INC.
1165 Chess Dr., Suite G
Foster City, CA 94404, USA
(650) 578-1396
(650) 578-8836
www.applikon.com

B. BRAUN BIOTECH, INC.
999 Postal Rd.
Allentown, PA 18103, USA
(800) 258-9000
(610) 266-9319
www.bbraunbiotech.com

BECKMAN COULTER, INC.
4300 N. Harbor Blvd.
Fullerton, CA 92834-3100, USA
(714) 871-4848
(714) 773-8898
www.beckman.com

BECTON DICKINSON LABWARE
Two Oak Park
Bedford, MA 01730, USA
(800) 343-2035
(617) 275-0043
www.bd.com/labware

Tissue Engineering. Essentials for Daily Laboratory Work W. W. Minuth, R. Strehl, K. Schumacher
Copyright © 2005 WILEY-VCH Verlag GmbH & Co. KGaA, Weinheim
ISBN: 3-527-31186-6

BEL-ART PRODUCTS
6 Industrial Rd.
Pequannock, NJ 07440-1992, USA
(973) 694-0500
(973) 694-7199
www.bel-art.com

BIO-RAD LABORATORIES, INC.
1000 Alfred Nobel Dr.
Hercules, CA 94547, USA
(510) 724-7000
(510) 741-1051
www.bio-rad.com

BIOCHROM KG
Leonorenstr. 2-6
D-12247 Berlin, Germany
+ 49 30 7799060
+ 49 30 7710012
www.biochrom.de

BIOCLONE AUSTRALIA PTY LTD.
54C Fitzroy St., Marrickville
Sydney, NSW Australia 2204
+ 61 2 517 1966
+ 61 2 517 2990
www.bioclone.com.au

BIOCON, INC.
15801 Crabbs Branch Way
Rockville, MD 20855, USA
(301) 417-0585
(301) 417-9238
www.bioconinc.com

BIODESIGN INC. OF NEW YORK
P.O. Box 1050
Carmel, NY 10512, USA
(845) 454-6610
(845) 454-6077
www.biodesignofny.com

BIOENGINEERING AG
Sagenrainstrasse 7
CH-8636 Wald, Switzerland
+ 41 55 256 8 111
+ 41 55 256 8 256
www.bioengineering.ch

BIOSOURCE INT'L INC.
Biofluids Division
1114 Taft St.
Rockville, MD 2085, USA
(301) 424-4140
(301) 424-3619
www.biofluids.com

BIOINVENT INT'L AB
Solvegatan 41
Lund, Sweden SE-223 70
+ 46 46 286 85 50
+ 46 46 211 08 06
www.bioinvent.com

BIOLOG LIFE SCIENCE INSTITUTE
Flughafendamm 9A
P.O. Box 107125
D-28071 Bremen, Germany
+ 49 421 591355
+ 49 421 5979713
www.biolog.de

BIOLOGICAL INDUSTRIES CO. LTD.
Kibbutz Beit Haemek
Israel 25115
+ 972 4 996 0595
+ 972 4 996 8896
www.bioind.com

BIOMEDICAL TECHNOLOGIES, INC.
378 Page St.
Stoughton, MA 02072, USA
(781) 344-9942
(781) 341-1451
www.btiinc.com

BIONIQUE TESTING
LABORATORIES, INC.
RR#1, Box 196, Fay Brook Drive
Saranac Lake, NY 12983, USA
(518) 891-2356
(518) 891-5753
www.bionique.com

BIORELIANCE
14920 Broschart Rd.
Rockville, MD 20850-3349, USA
(800) 553-5372
(301) 610-2590
www.bioreliance.com

BIOWHITTAKER, INC.
8830 Biggs Ford Rd.
Walkersville, MD 21793, USA
(301) 898-7025
(301) 845-8338
www.biowhittaker.com

BY-PROD CORP.
P.O. Box 66824
St. Louis, MO 63166, USA
(314) 534-3122
(314) 534-4422
www.bypcorp.com

CAMBIO
34 Newnham Rd.
Cambridge, U.K. CB3 9EY
+ 44 1223 366500
+ 44 1223 350069
www.cambio.co.uk

CELL WORKS INC.
University of Maryland
5202 Westland Blvd.
Baltimore, MD 21227, USA
(410) 455-5852
(410) 455-5851
www.cell-works.com

CELLEX BIOSCIENCES, INC.
8500 Evergreen Blvd.
Minneapolis, MN 55433, USA
(612) 786-0302
(612) 786-0915
www.cellexbio.com

CELLTECH GROUP plc
216 Bath Rd., Slough
Berkshire U.K. S11 9DL
+ 44 753 534655
+ 44 753 536632
www.celltechgroup.com

CELSIS INT'L PLC
Cambridge Science Park
Milton Rd.,
Cambridge, U.K. CB4 0FX
+ 44 01223 426008
+ 44 01223 426003
www.celsis.com

CHARLES RIVER TEKTAGEN
358 Technology Drive
Malvern, PA 19355, USA
(610) 640-4550
(610) 889-9028
www.tektagen.com

CLONETICS CELL SYSTEMS,
8830 Biggs Ford Rd.
Walkersville, MD 21793, USA
(301) 898-7025
(301) 845-8338
www.clonetics.com

COOK BIOTECH INC.
3055 Kent Avenue
W. Lafayette, IN 47906, USA
(765) 497-3355
www.cookgroup.com/cook_biotech

CORNING INC.
Science Products45
NAGOG PARK Division
Acton, MA 01720, USA
(978) 635-2200
(978) 635-2476
www.scienceproducts.corning.com

CSL LTD.
45 Poplar Rd., Parkville
Victoria Australia 3052
+ 61 3 93891389
+ 61 3 93891646
www.csl.com.au

CYMBUS BIOTECHNOLOGY LTD.
Unit J, Eagle Close, Chandlersford
Hampshire, U.K. S053 4NF
+ 44 8026 7676
+ 44 8026 7677
www.cymbus.co.uk

CYTOGEN RESEARCH AND
DEVELOPMENT, INC.
89 Bellevue Hill Rd.
West Roxbury, MA 02132, USA
(617) 325-7774
(617) 327-2405

CYTOVAX BIOTECHNOLOGIES INC.
8925 51 Avenue, Ste. 308
Edmonton, Alberta
Canada T6E 5J3
(780) 448-0621
(780) 448-0624
www.cytovax.com

DSM BIOLOGICS EUROPE
Zuiderweg 72/2, P.O. Box 454
Groningen, Netherlands 97+44 AP
+ 31 50 5222 222
+ 31 50 5222 333
www.dsmbiologics.com

EUROPEAN COLLECTION
OF CELL CULTURES
Porton Down, Salsbury
Wiltshire, U.K. SP4 OJG
+ 44 1980 612512
+ 44 1980 611315
www.ecacc.org

EXALPHA BIOLOGICALS, INC.
20 Hampden Street
Boston, MA 02119, USA
(617) 445-6463
(617) 989-0404
www.exalpha.com

EXOCELL, INC.
3508 Market Street, Suite 420
Philadelphia, PA 19104, USA
(215) 222-5515
(215) 222-5325
www.exocell.com

FORGENE, INC.
549 Eagle Street, P.O. Box 1370
Rhinelander, WI 54501, USA
(715) 369-8733
(715) 369 8737
www.insti-trees.com

GROPEP, LTD.
P.O. Box 10065, Gouger St.
Adelaide, South Australia 5000
618 8354 7709
618 8354 7777
www.gropep.com.au

HARLAN BIOPRODUCTS
FOR SCIENCE, INC.
P.O. Box 29176
Indianapolis, IN 46229, USA
(317) 359-1000
(317) 357-9000
www.hbps.com

HUMAN BIOLOGICS INT'L
7150 E. Camelback Rd., Suite 245
Scottsdale, AZ 85251, USA
(602) 990-2005
(602) 990-2155
www.humanbiologics.com

HYCLONE
LABORATORIES, INC.
1725 South HyClone Rd.
Logan, UT 84321, USA
(435) 753-4584
(435) 753-4589
www.hyclone.com

IDEXX LABORATORIES, INC.
One Idexx Dr.
Westbrook, ME 04092, USA
(207) 856-0300
(207) 856-0347
www.idexx.com

IGEN INT'L, INC.
16020 Industrial Dr.
Gaithersburg, MD 20877, USA
(301) 984-8000
(301) 208-3799
www.igen.com

IMCLONE SYSTEMS INC.
180 Varrick Street
New York, NY 10014, USA
(212) 645-1405
(212) 645-2054
www.imclone.com

IMMUNOVISION, INC.
1820 Ford Ave.
Springdale, AZ 72764
(800) 541-0960
www.immunovision.com

INFORS HT
Rittergasse 27, CH-4103 Bottmingen
Switzerland
+ 41 61 425 77 00
+ 41 61 425 77 01
www.infors.ch

INTERGEN CO. (EUROPE)
The Magdalen Centre
Oxford Science Park
Oxford, U.K. OX4 4GA
+ 44 1865 784647
+ 44 1865 784648

JOUAN, INC.
170 Marcel Dr.
Winchester, VA 22602, USA
(800) 662-7477
(540) 869-8626
(541)   www.jouan.com

JRH BIOSCIENCES
13804 West 107 St.
Lenexa, KS 66215, USA
(913) 469-5580
(913) 469-5584
www.jrhbio.com

KRAEBER GmbH & CO.
Waldhofstr. 14, D-25474 Ellerbek
Germany
+ 49 4101 30530
+ 49 4101 305390
www.kraeber.de

KENDRO LABORATORY
PRODUCTS
31 Pecks Lane
Newtown, CT 06470-2337, USA
(203) 840-6040
(203) 270-2210
www.kendro.de

LIFE TECHNOLOGIES, INC.
9800 Medical Center Way
Rockville, MD 20850, USA
(800) 828-6686
(800) 352-1468
www.lifetech.com

MATRITECH INC.
330 Nevada St.
Newton, MA 02460, USA
(617) 928-0820
(617) 928-0821
www.matritech.com

MEDAREX, INC.
1545 Route 22 East
Annandale, NJ 08801, USA
(908) 713-6001
(908) 713-6002
www.medarex.com

MEDICORP INC.
5800 Royalmount
Montreal, Quebec
Canada H4P 1K5
(514) 733-1900
(514) 733-1212
www.medicorp.com

MICRODYN TECHNOLOGIES, INC.
P.O. Box 98269
1204 Briar Patch Lane
Raleigh, NC 27624, USA
(919) 872-9375
(919) 872-9375
www.microdyn.de

MINUCELLS and MINUTISSUE GmbH
Starenstrasse 2
D – 93077 Bad Abbach, Germany
+49 (0) 9405 962440
+49 (0) 9405 962441
www.minucells.de

MOLECULAR PROBES INC.
4849 Pitchford Ave.
Eugene, Oregon 97402, USA
(541) 465-8300
(541) 344-6504
www.probes.com

NEW BRUNSWICK SCIENTIFIC
(U.K.) LTD.
163 Dixons Hill Rd.
North Mymms
Hatfield, Herts, U.K. AL9 7JE
+ 44 1707 275733
+ 44 1707 267859
www.nbsc.com

NEW BRUNSWICK SCIENTIFIC CO.,
INC.
P.O. Box 4005, 44 Talmadge Rd.
Edison, NJ 08818-4005, USA
(732) 287-1200
(732) 287-4222
www.nbsc.com

NEWPORT BIOSYSTEMS, INC.
1860 Trainor St.
Red Bluff, CA 96080, USA
(530) 529-2448
(530) 529-2648
www.newportbio.com

NORTHVIEW BIOSCIENCES, INC.
1880 Holste Rd.
Northbrook, IL 60062, USA
(847) 564-8181
(847) 564-8269
www.northviewlabs.com

NORTON PERFORMANCE PLASTICS
P.O. Box 3660
Akron, OH 44309-3660, USA
(216) 798-9240
(216) 798-0358
www.tygon.com

PAA LABORATORIES GmbH
Wiener Strasse 131
Linz, Upper Austria
Austria, A-4020
+ 43 732 33 08 90
+ 43 732 33 08 94
www.paa.at

PALL CORPORATION
2200 Northern Blvd.
East Hills, NY 11548, USA
(516) 484-5400
(516) 484-3637
www.pall.com

PHARMAKON
RESEARCH INT'L, INC.
P.O. Box 609
Waverly, PA 18471, USA
(717) 586-2411
(717) 586-3450
www.pharmakon.com

PROMOCELL BIOSCIENCE
ALIVE GmbH
Handschuhsheimer Landstr. 12
D-69120 Heidelberg, Germany
+ 49 6221 649340
+ 49 6221 6493440
www.promocell.com

Q-ONE BIOTECH LTD.
Todd Campus
West of Scotland Science Park
Glasgow, Scotland, U.K. G20 OXA
+ 44 141 946-9999
+ 44 141 946-0000
www.q-one.com

ROCKLAND IMMUNOCHEMICALS INC.
Box 316
Gilbertsville, PA 19525, USA
(610) 369-1008
(610) 367-7825
www.rockland-inc.com

SCHLEICHER & SCHUELL GmbH
Postfach 4
D-37582 Dassel, Germany
+ 49 5561 791 417
+ 49 5561 791 544
www.s-und-s.de

SEROLOGICALS CORP.
Fleming Road, Kirkton Campus
Livingston, U.K. EH54 7BN
+ 44 1506 404000
+ 44 1506 415210
www.serologicals.com

SEROTEC LTD.
22, Bankside, Station Approach,
Kidlington Oxford
U.K. OX5 IJE
+ 44 1865 852700
+ 44 1865 373899
www.serotec.co.uk

SIGMA CELL CULTURE
P.O. Box 14508
St. Louis, MO 63178, USA
(800) 521-8956
(314) 771-0633
www.sigma.com

SOLOHILL ENGINEERING INC.
4220 Varsity Dr.
Ann Arbor, MI 48108, USA
(313) 973-2956
(313) 973-3029
(314) www.solohill.com

SPECTRUM LABORATORIES, INC.
18617 Broadwick Street
Rancho Dominguez, CA 90220, USA
(310) 885-4600
(310) 885-4666
www.spectrumlabs.com

TCS CELLWORKS, LTD.
Botolph Claydon, Buckingham
Botolph Claydon, Buckingham
Bucks, U.K. MK18 2LR
+ 44 1296 71 3120
+ 44 1296 71 3122
www.tcscellworks.co.uk

TECHNE INC.
743 Alexander Rd.
Princeton, NJ 08540
(609) 452-9275
(609) 987-8177
www.techneusa.com

TEXAS BIOTECHNOLOGY CORP
7000 Fannin St., Suite 1920
Houston, TX 77030, USA
(713) 796-8822
www.tbc.com

WESTFALIA SEPARATOR AG
Werner-Habig-Str. 1
D-59302 Oelde, Germany
+ 49 2522 770
+ 49 2522 77 24 88
www.westfalia-separator.com

WHATMAN INC.
9 Bridewell Place
Clifton, NJ 07014, USA
(973) 773-5800
(973) 472-6949
www.whatman.com

WORTHINGTON BIOCHEMICAL
CORP.
730 Vassar Ave.
Lakewood, NJ 08701, USA
(732) 942-1660
(732) 942-9270
www.worthington-biochem.com

YES BIOTECH LABORATORIES LTD.
7035 Fir Tree Dr., Unit 23
Mississauga, Ontario
Canada L5S 1V6
(905) 677-9221
(905) 677-0023
www.yesbiotech.com

ZEPTOMETRIX CORP.
872 Main St.
Buffalo, NY 14202, USA
(716) 882-0920
(716) 882-0959
www.zeptometrix.com

Literature

Books

1. **Biomaterial Science:** An Introduction to materials in Medicine. Buddy D. Ratner et al., 2004, Academic Press
2. **Tissue Engineering** (Principles and Applications in Engineering), Bernhard Palsson, Jeffrey A. Hubell, Robert Plonsey, 2003, CRC Press.
3. **Principles of Tissue engineering**, Robert P. Lanza, Robert Langer, Joseph Vacanti (eds.), 2000, Academic Press.
4. **Frontiers in Tissue engineering**, Charles W. Patrick jr, Antonios G. Mikos, Larry V. McIntire (eds.). 1998, Pergamon.

Reviews and Reserach Papers

1. Gilbert TW, Stolz DB, Biancaniello F, Simons-Byrd A, Badylak SF. Production and character-ization of ECM powder: implications for tissue engineering applications. Biomaterials 26(12):1431-1435 (2005).
2. Datta NN, Holtorf HL, Sikavitsas VI, Jansen JA, Mikos AG. Effect of extracellular matrix synthesized in vitro on the osteoblastic differentiation of marrow stromal cells. Biomaterials 26(9):971-7 (2005).
3. Strehl R, Schumacher K, Minuth WW. Controlled respiratory gas delivery to embryonic renal epithelial explants in perfusion culture. Tissue Engineering 10(7):1196-1202 (2004).
4. Montanya E. Islet- and stem-cell-based tissue engineering in diabetes. Curr Opin Biotechnol 15(5):435-440 (2004).
5. Koh CJ, Atala A. Tissue engineering for urinary incontinence applications. Minerva Ginecol 56(4):371-8 (2004).
6. Minuth WW, Sorokin L, Schumacher K. Generation of renal tubules at the interface of an artificial interstitium. Cell Physiol Biochem 14(4-6):387-394 (2004).
7. Zammaretti P, Jaconi M. Cardiac tissue engineering: regeneration of the wounded heart. Curr Opin Biotechnol 15(5):430-4 (2004).
8. Pellegrini G. Changing the cell source in cell therapy. N England J Med 351(12):1170-2 (2004).
9. Sittinger M, Hutmacher DW, Risbud MV. Current strategies for cell delivery in cartilage and bone regeneration. Curr Opin Biotechnol 15(5):411-8 (2004).
10. Pradeep AR, Karthikeyan BV. Tissue engineering: prospect for regenerating periodontal tissue. Indian J Dent Res 14(4):224-9 (2004).
11. Levenberg S, Langer R. Advances in tissue engineering. Curr Top Dev Biol 61:113-34 (2004).
12. Rippon HJ, Bishop AE. Embryonic stem cells. Cell Prolif 37(1):23-34 (2004).
13. Langer R, Tirrell DA. Designing materials for biology and medicine. Nature 428(6982):487-92 (2004).

14. Minuth WW, Strehl R, Schumacher K. Tissue Factory – Conceptual design of a modular system for the in-vitro generation of functional tissues. Tissue Engineering 10,1/2:285-294 (2004).
15. Murray PE, Garcia-Godoy F. Stem cell responses in tooth regeneration. Stem Cells Dev 13(3):255-62 (2004).
16. Rocha FG, Wang EE. Intestinal tissue engineering: from regenerative medicine to model systems. J Surg Res 120(2):320-5 (2004).
17. Cowin SC. Tissue growth and modeling. Annu Rev Biomed Eng 6:77-107 (2004).
18. Gleason RL, Hu JJ, Humphrey JD. Building a functional artery: issues from the perspective of mechanics. Front Biosci 9:2045-55 (2004).
19. Qian L, Saltzman WM. Improving the expansion and neuronal differentiation of mesenchymal stem cells through culture surface modification. Biomaterials 25(7-8):1331-7 (2003).
20. Vogt AK, Lauer L, Knoll W, Offenhausser A. Micropatterned substrates for the growth of functional neuronal networks of defined geometry. Biotechnol Prog. 19(5):1562-8 (2003).
21. Malchesky PS. Artificial organs 2003: a year of review. Artif Organs 28(4):410-24 (2003).
22. Jahoda CA, Whitehouse J, Reynolds AJ, Hole N. Hair follicle dermal cells differentiate into adipogenic and osteogenic lineages. Exp Dermatol 12(6):849-59 (2003).
23. Nishida K. Tissue engineering of the cornea. Cornea 22(7),S28-34 (2003).
24. Schumacher K, Strehl R, Minuth WW. Urea restrains aldosterone-induced development of peanut agglutinin- binding on embryonic renal collecting duct epithelia. J Am Soc Nephrol 14:2758-2766 (2003).
25. Altman GH, Diaz F, Jakuba C, Calabro T, Horan RL, Chen J, Lu H, Richmond J, Kaplan DL. Silk-based biomaterials, Biomaterials 24,401-416 (2003).
26. Schumacher K, Strehl R, de Vries U, Minuth WW. Advanced technique for long term culture of epithelia in a continuous luminal - basal medium gradient. Biomaterials 23,3:805-815 (2002).
27. Kramer-Schultheiss KS, Schultheiss D. From wound healing to modern tissue engineering of the skin. A historical review on early techniques of cell and tissue culture. Hautarzt 53,751-760 (2002).
28. Warren SM, Sylvester K, Chen CM, Hedrick MH, Longaker MT. New directions in bioresorbable technology. Orthopedics 25,1201-1210 (2002).
29. Strehl R, Schumacher K, de Vries U, Minuth WW. Proliferating Cells versus Differentiated Cells in Tissue Engineering. Tissue Engineering 8,1:37-42 (2002).
30. Claes L, Ignatius A. Development of new, biodegradable implants. Chirurg 73:990-996 (2002).
31. Nomi M, Atala A, Coppi PD, Soker S. Principals of neovascularization for tissue engineering. Mol Aspects Med 23, 463 (2002).
32. Nasseri BA, Vacanti JP. Tissue engineering in the 21st century. Surg Technol Int 10,25-37 (2002).
33. Ushida T, Furukawa K, Toita K, Tateishi T. Three-dimensional seeding of chondrocytes encapsulated in collagen gel into PLLA scaffolds. Cell Transplant 11,489-494 (2002).
34. Jiang J, Kojima N, Kinoshita T, Miyamjia A, Yan W, Sakai Y. Cultivation of fetal liver cells in a three-dimensional poly-l-lactic acid scaffold in the presence of oncostatin M. Cell Transplant 11,403-406 (2002).
35. Atala A. Experimental and clinical experience with tissue engineering techniques for urethral construction. Urol Clin North Am 29,485-492 (2002).
36. Hinterhuber G, M;arquardt Y, Diem E, Rappersberger K, Wolff K, Foedinger D. Organotypic keratinocyte coculture using normal human serum: an immunomorphoöogical study at light and electron microscopic levels. Exp Dematol 11,413-420 (2002).
37. Chung T, Lu Y, Wang S, Lin Y, Chu S. Growth of human endothelial cells on photochemically grafted Gly-Arg-Gly-Asp (GRGD) chitosans. Biomaterials 23,4803 (2002).
38. van Luyn M, Tio R, Galleo y van Seijen X, Plantinga J, de Leij L, deJongste M, van Wachem P. Cardiac tissue engineering: characteristics of in unison contracting two- and three-dimensional neonatal rat ventricle cell (co)-cultures. Biomaterials 23,4793 (2002).
39. Sherwood J, Riley S, Palazzolo R, Brown S, Monkhouse D, Coates M, Griffith L, Landeen L, Ratcliff A. A three-dimensional osteochondral composite scaffold for articular cartilage repair. Biomaterials 23,4793 (2002).

40. Ozawa T, Mickle DA, weisel RD, Koyama N, Orawa S, Li RK. Optimal biomaterial for creation of autologous cardiac grafts. Circulation 106,176-182 (2002).
41. Cebotari S, Mertsching H, Kallenbach K, Kostin S, Repin O, Batrinac A, Kleczka C, Ciubotaru A, Haverich A. Construction of autologous human heart valves based on a acellular allograft matrix. Circulation 106,163-168 (2002).
42. Young CS, Terada S, Vacanti JP, Honda M, Bartlett JD, Yelick PC. Tissue engineering of complex tooth structures on biodegradable polymer scaffolds. J Dent Res 81,695-700 (2002).Landers R, Hubner U, Schmelzeisen R, Mulhaupt R. Rapid prototyping of scaffolds derived from thermoreversible hydrogels and tailored for applications in tissue engineering. Biomaterials 23,4437-4447 (2002).
43. Hori Y, Nakamura T, Kimura D, Kaino K, Kurokawa Y, Satomi S. Functional analysis of the tissue-engineered stomach wall. Artif Organs 26,868-872 (2002).
44. Bancroft Gn, Sikavitsas VI, van den Dolder J, Sheffiled TL, Ambrose CG, Jansen JA, Mikos AG. Fluid flow increases mineralitzed matrix deposition in 3D perfusion culture of marrow stromal osteoblastsa in a dose-dependent manner. Proc. Natl Acad Sci USA 99,12600 (2002).
45. Pei M, Solchaga LA, Seidel J, Zeng L, Vunjak-Novakovic G, Caplan AI, Freed LE. Bioreactors mediate the efficitiveness of tissue engineering scaffolds. FASEB J 16,1691-1694 (2002).
46. Burdick J, Mason M, Hinman A, Thorne K, Anseth K. Delivery of osteoinductive growth factors from degradable PEG hydrogels influences osteoblast differentiation and mineralization. J Control Release 83, 53 (2002).
47. Bucheler M, Wirz C, Schutz A, Bootz F. Tissue engineering of human salivary gland organoids. Acta Otolaryngol 122,541-545 (2002).
48. Grikscheidt TC, Ogilvie ER, Alsberg E, Mooney D, Vacanti JP. Tissue-engineered colon exhibits function in vivo. Surgery 132,200-204 (2002).
49. Alsberg E, Anderson KW, Albeiruti A, Rowley JA, Mooney DJ. Engineering growing tissues. Proc Natl Acad Sci USA 99,12025 (2002).
50. Wozney JM. Overview of bone morphogenic proteins. Spine 27,S2-S8 (2002).
51. Hollister SJ, Maddox RD, Taboas JM. Optimal design and fabrication of scaffolds to mimic tissue properties and satisfy biological strains. Biomaterials 23,4095 (2002).
52. Eschenhage T, Didie M, Heubach J, Ravens U, Zimmermann WH. Cardiac tissue engineering. Transpl Immunol 9,315-321 (2002).
53. Vats, Tolley NS, Polak JM, Buttery LD. Stem cells: sources and applications. Clin Otolaryngol 27,227-232 (2002).
54. Davisson T, Kunig S, Chen A, Sah R, Ratcliff A. Static and dynamic compression modulate matrix metabolism in tissue engineered cartilage. J Orthop Res 20,842-848 (2002).
55. Shimmura S, Tsubota K. Ocular surface reconstruction update. Curr Opin Ophthalmol 13,213-219 (2002).
56. Pascual G, Jurado F, Rodriguez M, Corrales C, Lopez-Hervas P, Bellon JM, Bujan J. The use of ischaemic vessels as protheses or tissue engineering scaffolds after cryopreservation. Eur J Vasc Endovasc Surg 24,23-30 (2002).
57. Risbud MV, Sittinger M. Tissue engineering : advances in in vitro cartilage generation. Trends Biotechnol 20,351-356 (2002).
58. Fansa H, Schneider W, Wolf G, Keilhoff G. Influence of insulin-like growth factor-I (IGF-I) on nerve autografts and tissue-engineered nerve grafts. Muscle Nerve 26,87-93 (2002).
59. Bishop AE, Buttery LD, Polak JM. Embryonic stem cells. J Pathol 197,424-429 (2002).
60. Haisch A, Klaring S, Groger A, Gebert C, Sittinger M. A tissue-engineered model for the manufacture of auricular-shaped cartilage implants. Eur Arch Otorhinolaryngol 259,316-321 (2002).
61. Hildebrandt KA, Jia F, Woo SL. Response of donar and recipient cells after transplantation of cells to the ligament and tendon. Microsc Res Tech 58,34-38 (2002).
62. Cheng B, chen Z. Fabricating autologous tissue to engineer artificial nerve. Microsurgery 22,133-137 (2002).

Subject Index

a

actin 27, 117
action potential 245
activin 61, 68
adaptation 72, 73
adenylate cyclase 244
adherence 196
adhesion 41, 172
adipoblast 141
adipocyte 20
adipose tissue 20, 141
aggrecan 40
aldosteron 244
antibiotics 90, 91
antibody 255, 258
– monoclonal 95, 96
antigen expression 230
antimycotics 91
apoptosis 16, 43, 69–72, 75, 76
aquaporin 248
astrocyte 32
ATP (adenosine triphosphate) 6
atrophy 72
autologous system 131
axon 31

b

barrier
– continuity 188–190
– epithelial function 185
basal cell 16
– hyperplasia 73
biodegradation 176
biomodule 126
biophysical factors 206–207
bioreactor 126
– rotating 167
blood vessel 120, 124
BM-40 53

BMP

BMP (bone morphogenic protein) 69, 92
bone 17, 21–25
– construct 121
– lamellar 25
– regeneration 134
– repair 121
– replacement 22
branching morphogenesis 112
BSE 144
burns, large-scale 116, 117

c

cadherin 49
CAM (cell-adhesion molecule) 49
canaliculus 24
capillary
– module 168
– network 168
cardiomyocyte 29, 118
– cultivation 101–103
cartilage 17, 21–25, 65, 134, 228
– articular, defects 115
– capsule 22
– construct 121
– elastic 22, 23
– fibrous 23
– hyaline 22, 23, 105, 115
– joint 22
– territory 22
– types 22
cartilage oligomeric matrix protein (COMP)
   43
cell
– binding 44–46
– cytoskeleton 8, 241, 242
– ECM 8, 9, 19, 35–53
– endoplasmatic reticulum (ER) 6, 7
– endosome 7, 8
– functional exceptions 60

Tissue Engineering. Essentials for Daily Laboratory Work W. W. Minuth, R. Strehl, K. Schumacher
Copyright © 2005 WILEY-VCH Verlag GmbH Co. KGaA, Weinheim
ISBN: 3-527-31186-6